SpringerBriefs in Mathematics

SpringerBriefs in Mathematics showcases expositions in all areas of mathematics and applied mathematics. Manuscripts presenting new results or a single new result in a classical field, new field, or an emerging topic, applications, or bridges between new results and already published works, are encouraged. The series is intended for mathematicians and applied mathematicians.

More information about this series at http://www.springer.com/series/10030

Szymon M. Walczak

Metric Diffusion Along
Foliations

Springer

Szymon M. Walczak
National Science Center
Kraków, Poland

Faculty of Mathematics
  and Computer Science
University of Łódź
Łódź, Poland

ISSN 2191-8198   ISSN 2191-8201   (electronic)
SpringerBriefs in Mathematics
ISBN 978-3-319-57516-2·   ISBN 978-3-319-57517-9   (eBook)
DOI 10.1007/978-3-319-57517-9

Library of Congress Control Number: 2017939353

Mathematics Subject Classification: 53C12, 53C23

Printed on acid-free paper

This Springer imprint is published by Springer Nature
The registered company is Springer International Publishing AG
The registered company address is: Gewerbestrasse 11, 6330 Cham, Switzerland

To Joanna, Jan, Julia & Zuzanna

Contents

Introduction

In the last few years, the author of this book was looking for a kind of a deformation of a metric on a foliated Riemannian manifold, which will preserve, in some sense, the set of compact leaves. More precisely, given a foliation $\mathscr{F}$ on a compact Riemannian manifold (M, g), we are looking for a deformation D_t of the structure g or induced Riemannian metric d such that the limit of the set $\mathscr{C}$ of compact leaves under, for example, the Gromov–Hausdorff convergence is homeomorphic to $\mathscr{C}/\mathscr{F}$.

The first approach to this problem by warped foliations [22, 23] wasn't successful. Briefly speaking, for a compact foliated Riemannian manifold $(M, \mathscr{F}, g)$, and a smooth function $f : M \to (0, \infty)$ constant along the leaves of $\mathscr{F}$ the metric induced by the Riemannian structure g_f defined by

$$g_f(v, w) = f^2 g(v, w) \quad \text{for } v, w \text{ tangent to } \mathscr{F},$$

$$g_f(v, w) = g(v, w) \quad \text{if at least one of } v, w \text{ is perpendicular to } \mathscr{F}$$

is called the *warped metric*. f is called the *warping function*, while the metric space (M, d_f), where d_f is a metric induced by g_f, the *warped foliation*.

Let f_n be a sequence of warping functions converging uniformly to zero on a Reeb foliation $\mathscr{R}$ of an annulus $A = \{x \in \mathbb{R}^2 : 1 \leq \|x\| \leq 2\}$. In [23], the following was proved:

Theorem I.1 *The limit of warped Reeb foliation* $\lim_{n \to \infty}^{\text{GH}} (M, d_{f_n})_{n \in \mathbb{N}}$ *under the Gromov–Hausdorff convergence is a singleton.*

The above example shows that the two boundary compact leaves, which are linked by a non-compact leaf, collapse, while warping, to the same point of the limit. The same can be observed for compact foliations. In [22] it was shown that the non-empty bad set of the compact foliation described by Epstein and Vogt in [10] collapses, in Gromov–Hausdorff topology, to the singleton.

The approach presented here uses more advanced tools. First, observe that the Wasserstein distance d_W (see [19]) of two Dirac measures on a Polish metric space (X, d) concentrated in $x, y \in X$ is equal to $d(x, y)$. Having d_W, with foliated heat diffusion operators D_t (introduced by L. Garnett in [11]) we define, for given time $t \geq 0$, the *metric* $D_t d$ *diffused along a foliation* on a Riemannian manifold M equipped with a foliation $\mathscr{F}$ as the Wasserstein distance of Dirac measures diffused at time t. It occurs that $D_t d$ defines the same topology on M as the initial one (Theorem 4.1).

In further considerations, we concentrate our attention on compact foliations. The reason is topological, namely, that the leaves have no ends, so that the nature of the heat kernel is well known. Notice that in the case of the manifolds with ends in general there is no knowledge on the heat kernel behaviour. On the other hand, the existence of a non-empty bad set can produce a number of problems on convergence of the family $(M, D_t d)$.

The main purpose for this work is to try to answer the question, whether the family $(M, D_t d)$ converges or not in d_{WH} to a closed subset of $\mathscr{P}(M)$. This will be the main subject of the studies presented here.

There is one more problem to settle. In the case of warped foliations, the Gromov–Hausdorff convergence was used. In the case of metric diffusion something else is more appropriate.

Let us denote by $\mathscr{P}(M)$ the set of all Borel probability measures on M. There is a natural isometric embedding of $(M, D_t d)$ into $(\mathscr{P}(M), d_W)$ defined by

$$\iota_t : M \ni x \mapsto D_t \delta_x \in \mathscr{P}(M).$$

Hence, for $t, s \geq 0$ we can consider $(M, D_t d)$ and $(M, D_s d)$ as the closed subsets of $\mathscr{P}(M)$ (which is compact if M is so), and we shall use the Hausdorff distance of $\iota_t(M)$ and $\iota_s(M)$ in $(\mathscr{P}(M), d_W)$. With the above in mind, we will write $(M, D_t d)$ or simply M_t instead of $(\iota_t(M), d_W)$.

The book was planned to provide all necessary facts needed to understand the metric diffusion along compact foliations, that is some basic facts from the optimal transportation theory and the theory of foliations. Chapter 1 is devoted to the Wasserstein distance, Kantorovich Duality Theorem, and the metrization of the weak-* topology by the Wasserstein distance. Moreover, we prove some technical lemmas used in further considerations. In Chapter 2, we present some basics about foliations, holonomy, and heat diffusion. They are necessary to understand the notion of the metric diffusion. The compact foliations are discussed in Chapter 3 where we recall these facts which are essential for further considerations.

The main results are presented in Chapter 4. We define the metric diffusion $D_t d$ and study the topology of the metric space $(M, D_t d)$. The remaining pages are devoted to the limits of diffused metrics along compact foliations. We prove the necessary conditions for Wasserstein–Hausdorff convergence of the metric diffused along compact foliation with non-empty Epstein hierarchy. The first result (Theorem 4.6) provides an information about the geometry of the compact foliation, that is it describes the leaf volume growth near connected components of the bad sets. The second (Theorem 4.7) is rather measure-theoretic one. Enhancement of the necessary conditions presented in Theorem 4.6 and Theorem 4.7 allows us to formulate the sufficient condition of Wasserstein–Hausdorff convergence of metrics diffused along compact foliations of dimension one with finite Epstein hierarchy.

As a kind of supplement, we present some facts about the metric diffusion along non-compact foliations. We provide the full description of the limit for metrics diffused along foliation with at least one compact leaf on the two-dimensional torus T^2.

I would like to express my thanks to Prof. Jesus A. Alvarez-Lopez from the University of Santiago de Compostela for the initial idea of metric diffusion and a number of fruitful discussions. Thanks are also due to Prof. Paweł Walczak and my colleagues from the University of Łódź, namely Wojciech Kozłowski, Kamil Niedziałomski, and Krzysztof Andrzejewski for helpful discussions and important remarks, and to Andrzej Komisarski for some probabilistic explanations. I'm also grateful to Prof. Takashi Tsuboi from the University of Tokyo, who gave me the opportunity to spend some time in Tokyo, where this book has got the current shape. Last but not least, I would like to acknowledge my gratitude to the University of Łódź, Prof. Ryszard Pawlak, the Dean of the Faculty of Mathematics and Computer Science, and to the National Center of Science (NCN, grant # 6065/B/H03/2011/40) for financial support during the research on metric diffusion.

Chapter 1
Wasserstein Distance

The Wasserstein distance of Borel probability measures plays a very important role in metric diffusion along foliations. In this chapter we present some foundations of the Optimal Transportation Theory, that is, the Kantorovich Duality Theorem for optimal transportation problem and the definition of the Wasserstein distance, together with the weak-$*$ topology metrization theorem for the set $\mathscr{P}(X)$ of Borel probability measures on a compact metric space X. The chapter is closed by some technical lemmas used in the later considerations.

The full theory of the Optimal Transportation can be found in the excellent monographs by C. Villani [18, 19]. We only present the main aspects, which are needed to understand the theory of metric diffusion. Full description and precise proofs can be found in the monographs mentioned above.

If not mentioned, we will always assume that topological spaces, metric spaces or manifolds we consider are compact. For full generality, refer to [18] and [19].

1.1 Optimal Transportation Problem

The Wasserstein distance of measures on compact metric spaces comes from the Monge–Kantorovich problem of existence of an optimal transportation plan. In other words, imagine that you have a pile of *something* which should fill perfectly a hole in the ground. Of course, both the pile and the hole have the same volume. There are many ways of doing this, but the one we are interested in is that with the lowest cost. *The cost* can be understood as an effort needed to do the job.

We now restrict our attention to Borel probability measures μ and ν on compact spaces X and X', respectively. A Borel probability measure π on $X \times X'$ is called a *coupling with marginals μ and ν* (Figure 1.1) if it satisfies

$$\pi(A \times X') = \mu(A) \text{ and } \pi(X \times B) = \mu(B).$$

© The Author(s) 2017
S.M. Walczak, *Metric Diffusion Along Foliations*, SpringerBriefs in Mathematics,
DOI 10.1007/978-3-319-57517-9_1

Fig. 1.1 A coupling

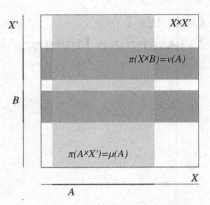

The set of all couplings with marginals μ and ν will be denoted by $\Pi(\mu, \nu)$. Note that $\Pi(\mu, \nu)$ is non-empty, since the product measure $\mu \times \nu \in \Pi(\mu, \nu)$.

Let $c : X \times X' \to \mathbb{R} \cup \{+\infty\}$ be a measurable map. c is called *the cost function*, and it defines the cost of transportation of unit mass from $x \in X$ to $x' \in X'$. The Kantorovich Optimal Transportation Problem, which lies under the concept of the Wasserstein distance can be formulated as follows:

$$\text{Minimize } I(\pi) = \int_{X \times X'} c(x, y) d\pi(x, y) \text{ for all } \pi \in \Pi(\mu, \nu).$$

Define

$$J(\varphi, \psi) = \int_X \varphi(x) d\mu(x) + \int_{X'} \psi(y) d\nu(y),$$

where $\varphi \in L^1(d\mu)$ and $\psi \in L^1(d\nu)$. Denote by Φ_c the set of all measurable functions $(\varphi, \psi) \in L^1(d\mu) \times L^1(d\nu)$ satisfying

$$\varphi(x) + \psi(x') \leq c(x, x').$$

for $d\mu$-almost all $x \in X$ and $d\nu$-almost all $x' \in X'$.

Theorem 1.1 (Kantorovich Duality Theorem)

$$\inf_{\pi \in \Pi(\mu, \nu)} I(\pi) = \sup_{(\varphi, \psi) \in \Phi_c} J(\varphi, \psi).$$

Moreover, the infimum on the left-hand side is attained.

Following explanations provided in [18], the key of the proof is to rewrite the constrained infimum problem as an "inf sup" problem and, applying minimax principle, replace an "inf sup" by "sup inf." The rigorous description of this proof is provided by Villani in [18]. We here repeat only a formal proof, also presented in [18].

Proof Write

$$\inf_{\pi \in \Pi(\mu,\nu)} I(\pi) = \inf_{\pi \in M_+(X \times X')} (I(\pi) + \mathrm{ind}(\pi)),$$

where $M_+(X \times X')$ denotes the space on nonnegative Borel measures on $X \times X'$, and ind denotes *the indicator function*, i.e., $\mathrm{ind}(\pi) = 0$ if $\pi \in \Pi(\mu, \nu)$, and $+\infty$ otherwise. Since the constraints defining Π are linear, one can write:

$$\mathrm{ind}(\pi) = \sup_{(\varphi,\psi)} \left(\int_X \varphi(x)d\mu(x) + \int_{X'} \psi(y)d\nu(y) \right.$$

$$\left. - \int_{X \times X'} (\varphi(x) + \psi(y))d\pi(x,y) \right)$$

where the supremum is over all $(\varphi, \psi) \in C(X) \times C(X')$, and $C(Y)$ denotes the set of continuous functions on Y. Thus

$$\inf_{\pi \in \Pi(\mu,\nu)} I(\pi) = \inf_{\pi \in M_+(X \times X')} \sup_{(\varphi,\psi)} \left(\int_{X \times X'} c(x,y)d\pi(x,y) \right.$$

$$+ \int_X \varphi(x)d\mu(x) + \int_{X'} \psi(y)d\nu(y)$$

$$\left. - \int_{X \times X'} (\varphi(x) + \psi(y))d\pi(x,y) \right).$$

Invoking the minimax principle, we rewrite the above as

$$\sup_{(\varphi,\psi)} \inf_{\pi \in M_+(X \times X')} \left(\int_{X \times X'} c(x,y)d\pi(x,y) \right.$$

$$+ \int_X \varphi(x)d\mu(x) + \int_{X'} \psi(y)d\nu(y)$$

$$\left. - \int_{X \times X'} (\varphi(x) + \psi(y))d\pi(x,y) \right)$$

$$= \sup_{(\varphi,\psi)} \left(\int_X \varphi(x)d\mu(x) + \int_{X'} \psi(y)d\nu(y) \right.$$

$$\left. - \sup_{\pi \in M_+(X \times X')} \int_{X \times X'} (\varphi(x) + \psi(y) - c(x,y))d\pi(x,y) \right).$$

Observe that if the function $\xi(x,y) = \varphi(x) + \psi(y) - c(x,y)$ takes a positive value at some point $(x_0, y_0) \in X \times X'$, then by choosing $\pi = \lambda \delta_{(x_0,y_0)}$, where $\delta_{(w,z)}$ denotes the Dirac mass at (w, z), and letting $\lambda \to \infty$, the supremum is infinite. On the other hand, if ξ is nonpositive $d\mu \otimes d\nu$-everywhere, the supremum is obtained for $\pi = 0$. Hence,

$$\sup_{\pi \in M_+(X \times X')} \int_{X \times X'} (\varphi(x) + \psi(y) - c(x,y)) d\pi(x,y) = \begin{cases} 0 & \text{if } (\varphi,\psi) \in \Phi_c, \\ +\infty & \text{otherwise.} \end{cases}$$

Finally,

$$\sup_{(\varphi,\psi)} \left(\int_X \varphi(x) d\mu(x) + \int_{X'} \psi(y) d\nu(y) \right.$$

$$- \sup_{\pi \in M_+(X \times X')} \int_{X \times X'} (\varphi(x) + \psi(y) - c(x,y)) d\pi(x,y))$$

$$= \sup_{(\varphi,\psi) \in \Phi_c} J(\varphi, \psi).$$

To complete the proof, it only remains to show that $\inf_{\pi \in \Pi(\mu,\nu)} I(\pi)$ is attained. This is a direct consequence of the compactness of $\Pi(\mu, \nu)$.

To begin, let $(\pi_k)_{k \in \mathbb{N}}$ be a minimizing sequence for I, and let π_* be any weak cluster point of $(\pi_k)_{k \in \mathbb{N}}$. Denote by $(c_n)_{n \in \mathbb{N}}$ the non-decreasing sequence of nonnegative, uniformly continuous cost functions such that $c = \sup_{n \in \mathbb{N}} c_n$. By the detailed proof of Kantorovich Duality Theorem provided in [18],

$$I_n(\pi) = \inf_{\pi \in \Pi(\mu,\nu)} \int_{X \times X'} c_n(x,y) d\pi(x,y) = \sup_{(\varphi,\psi) \in \Phi_{c_n}} J(\varphi, \psi)$$

and

$$\inf_{\pi \in \Pi(\mu,\nu)} I(\pi) = \sup_n I_n(\pi).$$

By monotone convergence of the sequence $(c_n)_{n \in \mathbb{N}}$,

$$I(\pi_*) = \lim_{n \to \infty} I_n(\pi_*) \le \lim_{n \to \infty} \limsup_{k \to \infty} I_n(\pi_k)$$

$$\le \limsup_{k \to \infty} I(\pi_k) = \inf_{\pi \in \Pi(\mu,\nu)} I(\pi).$$

This completes the proof. $\square$

1.2 Wasserstein Distance

Let (X, d) be a compact metric space. Denote by $\mathscr{P}(X)$ the set of all Borel probability measures on X. For $\mu, \nu \in \mathscr{P}(X)$ and any positive real number $p > 0$ define, following [19], the *associated optimal transportation cost operator* $\mathscr{T}_p$ by

$$\mathscr{T}_p(\mu, \nu) := \inf_{\Pi(\mu,\nu)} \int_{X \times X} d(x, y)^p d\pi(x, y),$$

Notice that the infimum in the above equality is attained in at least one coupling π_0 (due to Kantorovich Duality Theorem). Such a coupling is called *optimal*.

Let us now consider the set of all Borel probability measures $\mathscr{P}(X)$ on a compact metric space (X, d). Following [19], Theorem 7.3, we can formulate the following:

Theorem 1.2 *The formula*

$$d_{\mathrm{W},p}(\mu, \nu) = \begin{cases} \mathscr{T}_p^{\frac{1}{p}}(\mu, \nu) \text{ for } p \in (1, \infty], \\ \mathscr{T}_p(\mu, \nu) \text{ for } p \in (0, 1], \end{cases}$$

defines a metric on $\mathscr{P}(X)$.

Before we present a proof, we shall prove the following lemma.

Lemma 1.1 (Gluing Lemma) *For any probability measures $\mu_1, \mu_2, \mu_3 \in \mathscr{P}(X)$ denote by ρ_{12} and ρ_{23} couplings of μ_1, μ_2 and μ_2, μ_3, respectively. There exists a coupling ρ on $X \times X \times X$ with marginals ρ_{12} and ρ_{23}.*

Proof Let ρ_{12} and ρ_{23} couplings of μ_1, μ_2 and μ_2, μ_3, respectively. The disintegration theorem [12] allows us to write a probability measure on $X \times X$ as an average of probability measures on $\{x\} \times X$, $x \in X$. In particular, if π is a probability measure on $X \times X$, with marginal μ on the first factor X, then there exists a measurable application $X \ni x \mapsto \pi_x \in \mathscr{P}(X)$, uniquely determined $d\mu(x)$-almost everywhere, such that

$$\pi = \int_X (\delta_x \otimes \pi_x) d\mu(x),$$

that is, if $u \in C(X \times X)$ then

$$\int_{X \times X} u(x, u) d\pi(x, y) = \int_X (\int_X u(x, y) d\pi_x(y)) d\mu(x).$$

So, we can write

$$\rho_{12} = \int_X \rho_{12,2} \otimes \delta_{x_2} d\mu_2(x_2) \text{ and } \rho_{23} = \int_X \delta_{x_2} \otimes \rho_{23,2} d\mu_2(x_2).$$

We define $\rho \in \mathscr{P}(X \times X \times X)$ setting $\rho = \int_X \rho_{12,2} \otimes \delta_{x_2} \otimes \rho_{23,2} d\mu_2(x_2)$. ρ is our desired measure. □

Let us now return to the proof of Theorem 1.2.

Proof (of Theorem 1.2) It is obvious that $d_{\mathrm{W},p}$ is nonnegative, symmetric, and $d_{\mathrm{W},p}(\mu, \mu) = 0$ for any $\mu \in \mathscr{P}(X)$.

Let $d_{\mathrm{W},p}(\mu, \nu) = 0$. There exists a coupling ρ_0 concentrated on the diagonal $\mathrm{diag}(X \times X) = \{(x, x) : x \in X\}$. Hence, for any bounded function φ with compact support

$$\int_X \varphi(x) d\mu(x) = \int_{X \times X} \varphi(x) d\rho_0(x, y) = \int_{X \times X} \varphi(y) d\rho_0(x, y) = \int_X \varphi(y) d\nu(y).$$

Hence $\mu = \nu$.

Let us now suppose that ρ_{ij} are optimal couplings of μ_i and μ_j, $i, j \in \{1, 2, 3\}$. Let ρ be a measure constructed in Lemma 1.1. For $p \leq 1$, we have

$$\begin{aligned} d_{\mathrm{W},p}(\mu_1, \mu_3) &\leq \int_{X \times X} d(x_1, x_3) d\rho_{13}(x_1, x_3) = \int_{X \times X \times X} d(x_1, x_3) d\rho(x_1, x_3) \\ &\leq \int_{X \times X \times X} (d(x_1, x_2) + d(x_2, x_3)) d\rho(x_1, x_2, x_3) \\ &\leq \int_{X \times X} d(x_1, x_2) d\rho_{12}(x_1, x_2) + \int_{X \times X} d(x_2, x_3) d\rho_{23}(x_2, x_3) \\ &= d_{\mathrm{W},p}(\mu_1, \mu_2) + d_{\mathrm{W},p}(\mu_2, \mu_3). \end{aligned}$$

For $p > 1$ one should also use Minkowski's inequality. $\square$

The following theorem describes the magnificence of the Wasserstein distance (the theorem in the full generality can be found in [19], Theorem 7.12):

Theorem 1.3 *Let $(\mu_k)_{k \in \mathbb{N}}$ be a sequence of Borel probability measures on a compact metric space (X, d), $p \geq 1$. If $\mu \in \mathscr{P}(X)$, then the following statements are equivalent:*

1. *$d_{\mathrm{W},p}(\mu_i, \mu) \underset{i \to \infty}{\longrightarrow} 0$,*
2. *$\mu_i \underset{i \to \infty}{\longrightarrow} \mu$ in the weak sense.*

Proof First, let us suppose that $d_{\mathrm{W},p}(\mu_i, \mu) \underset{i \to \infty}{\longrightarrow} 0$. Since X is compact, the set $\{\mu_k\}_{k \in \mathbb{N}}$ is tight, and by Prokhorov Theorem (see [1]), there exists a subsequence $(\mu_{k'})$ converging in the weak sense to a measure $\tilde{\mu} \in \mathscr{P}(X)$.

By the lower semi-continuity of d^p,

$$d_{\mathrm{W},p}(\mu, \tilde{\mu}) \leq \lim_{k' \to \infty} \inf d_{\mathrm{W},p}(\mu_{k'}, \mu) = 0.$$

Hence, $\tilde{\mu} = \mu$.

Conversely, assume that $(\mu_k)_{k \in \mathbb{N}}$ converges weakly in $\mathscr{P}(X)$ to μ. Let us denote by π_k the optimal couplings of μ_k and μ such that

$$\int_{X \times X} d^p(x, y) d\pi_k(x, y) \xrightarrow[k \to \infty]{} 0.$$

Since X is compact and $(\mu_k)_{k \in \mathbb{N}}$ is tight, so is μ. Moreover, by Prokhorov Theorem applied to the sequence $(\pi_k)_{k \in \mathbb{N}}$, one may assume that $\pi_k \to \pi$. By the stability of optimal couplings [18, Theorem 5.20], π is an optimal coupling of μ and μ, so it is a trivial one. This ends our proof. □

In further considerations we shall use $d_{W,1}$, which will be shortly denoted by d_W.

One should notice that the Wasserstein distance $d_W(\delta_x, \delta_y)$ of Dirac masses concentrated in $x, y \in M$ coincides with the original distance $d(x, y)$. This follows directly from the fact that $\delta_{(x,y)}$ is the only coupling of δ_x and δ_y.

Let $\Delta^k = \{(t_1, \ldots, t_k) \in \mathbb{R}^k : \forall_{j \in \{1, \ldots, k\}} t_j \geq 0, \sum_j t_j = 1\}$.

Lemma 1.2 *The set*

$$\mathscr{D}(X) = \{\mu \in \mathscr{P}(X) : \mu = \sum_{i=1}^{k} t_i \delta_{x_i}, \ (t_1, \ldots, t_k) \in \Delta^k, \ x_1, \ldots, x_k \in X, k \in \mathbb{N}\}$$

is dense in $\mathscr{P}(X)$.

Proof Let $\mu \in \mathscr{P}(X)$ and $\varepsilon > 0$. Let us choose a partition $\{A_1, \ldots, A_k\}$ of X by Borel sets of diameter smaller than 2ε and points $x_i \in A_i$. Set $t_i = \mu(A_i)$. Since μ is a probability measure, then $t = (t_1, \ldots, t_k) \in \Delta^k$. Let $\nu = \sum_i t_i \delta_{x_i}$ and $\pi_\varepsilon(E) = \mu(\mathrm{pr}_1(E \cap (\bigcup_{i=1}^{k}(A_i \times \{x_i\}))))$, where pr_1 denotes the projection onto the first factor. π_ε is a coupling of μ and ν. Indeed, let $A \subset X$. Then

$$\pi_\varepsilon(A \times X) = \sigma(\mathrm{pr}_1((A \times X) \cap (\bigcup A_i \times \{x_i\}))) =$$

$$\sigma(\mathrm{pr}_1(\bigcup(A \cap A_i) \times \{x_i\}))) = \sigma(\bigcup(A \cap A_i)) = \sigma(A).$$

Moreover, for any x_i,

$$\pi_\varepsilon(X \times \{x_i\}) = \sigma(\mathrm{pr}_1(A_i \times \{x_i\})) = \sigma(A_i) = t_i = \nu(x_i).$$

Finally,

$$\int_{X \times X} d(x, y) d\pi_\varepsilon(x, y) = \int_{\bigcup(A_i \times \{x_i\})} d(x, y) d\pi_\varepsilon(x, y) \leq \sum t_i \mathrm{diam} A_i$$

$$\leq 2\varepsilon \sum t_i = 2\varepsilon.$$

The conclusion follows from the definition of d_W. □

Our next goal is to find some direct estimates for the Wasserstein distance of probability measures on a compact metric X. We will use them as a main technical tool in Chapter 4.

Let $\varepsilon \in (0,1)$ and $x_1, \ldots, x_k, y_1, \ldots, y_k \in X$. Suppose that $d(x_i, y_i) \leq \varepsilon$ for any $i \in \{1 \ldots, k\}$. Let

$$\mu = \sum_{i=1}^{k} t_i \delta_{x_i}, \quad \nu = \sum_{j=1}^{k} s_j \delta_{y_j}.$$

Obviously, $\mu, \nu \in \mathscr{D}(X)$.

Lemma 1.3 (Wasserstein–Gromov) *If $\sum_{i=1}^{k} |t_i - s_i| \leq \varepsilon$, then there exists a constant $C > 0$ depending only on (X, d) such that*

$$d_W(\mu, \nu) \leq C\varepsilon.$$

Proof Notice that the Wasserstein distance of measures $\mu = \sum_{i=1}^{k} t_i \delta_{x_i}$ and $\tilde{\nu} = \sum_{i=1}^{k} t_i \delta_{y_i}$ is estimated by $\sup_i d(x_i, y_i)$. Indeed, $\pi = \sum_{i=1}^{k} t_i \delta_{(x_i, y_i)}$ is a coupling of μ and $\tilde{\nu}$, and

$$d_W(\mu, \tilde{\nu}) \leq \int_{X \times X} d(x, y) d\pi(x, y) \leq \sum_{i=1}^{k} t_i d(x_i, y_i)$$

$$\leq \sup_{i \leq k} d(x_i, y_i) \sum_{i=1}^{k} t_i = \sup_{i \leq k} d(x_i, y_i).$$

The above estimation allows us to consider only the measures supported on the same set $\{x_1, \ldots, x_k\}$. Let $\pi_\mu = \sum_{i=1}^{k} a_i \delta_{x_i} + \sum_{j=1}^{k} r_j^\mu \delta_{y_j}$ where $a_i = \min\{t_i, s_i\}$, $i \in \{1, \ldots, k\}$, and $r_i^\mu = 0$ if $t_i \leq s_i$ and $t_i - s_i$ otherwise. Let $m_\mu = \sum_{i=1}^{k} a_i \delta_{(x_i, x_i)} + \sum_{j=1}^{k} r_j^\mu \delta_{(x_j, y_j)}$. Since $\sum_{i=1}^{k} r_j^\mu = \sum_{i=1}^{k} |t_i - s_i| \leq \varepsilon$, then

$$d_W(\mu, \pi_\mu) \leq \int_{X \times X} d(x, y) dm_\mu(x, y) \leq \sum_{i=1}^{k} r_j^\mu d(x_i, x_j) \leq \operatorname{diam}(X) \cdot \varepsilon.$$

Similarly, defining $\pi_\nu = \sum_{i=1}^{k} a_i \delta_{x_i} + \sum_{j=1}^{k} r_j^\nu \delta_{y_j}$ with $r_j^\nu = 0$ if $s_i \leq t_i$ and $s_i - t_i$ otherwise, we show that $d_W(\mu, \pi_\mu) \leq \operatorname{diam}(X) \cdot \varepsilon$.

Let $m_{\mu, \nu} = \sum_{i=1}^{k} a_i \delta_{(x_i, x_i)} + \sum_{j,l=1}^{k} r_j^\mu r_l^\nu \delta_{(y_i, y_j)}$. $m_{\mu, \nu}$ is a coupling of π_μ and π_ν. We have

$$d_W(\pi_\mu, \pi_\nu) \leq \int_{X \times X} d(x, y) dm_{\mu, \nu}(x, y) \leq \sum_{j,l=1}^{k} r_j^\mu r_l^\nu d(y_j, y_l)$$

$$\leq \operatorname{diam}(X) \sum_{j=1}^{k} r_j^\mu \sum_{l=1}^{k} r_l^\nu \leq \operatorname{diam}(X) \cdot \varepsilon^2.$$

Finally, returning to the initial measures, we get

$$d_W(\mu, \nu) \leq d_W(\mu, \pi_\mu) + d_W(\pi_\mu, \pi_\nu) + d_W(\pi_\nu, \nu)$$

$$\leq \text{diam}(X)(2\varepsilon + \varepsilon^2) + 2\varepsilon \leq C\varepsilon$$

with $C = 3\text{diam}(X) + 2$. This ends the proof. $\qquad\square$

Let $\varepsilon > 0$, and let $\mu \in \mathscr{P}(X)$. Consider $A \subset X$ such that $\mu(A) > 1 - \varepsilon$. Let $y_0 \in X \setminus A$, and let $\{U_1, \ldots, U_k\}$ be a partition of A by measurable sets of diameter smaller than ε, i.e.

1. $\bigcup_{i=1}^k U_i = A$,
2. $U_i \cap U_j = \emptyset$
3. $\text{diam}(U_j) \leq \varepsilon$ for any $1 \leq j \leq k$.

Let $U_0 = X \setminus A$. Choose points $y_j \in U_j$ and define a measure ν as

$$\nu = \sum \mu(U_j)\delta_{y_j} + (1 - \mu(U_0))\delta_{y_0}.$$

Lemma 1.4 $d_W(\mu, \nu) \leq (\text{diam}(X) + 1)\varepsilon$.

Proof Define a measure π on $X \times X$ by the formula

$$\pi(E) = \mu(\text{pr}_1(E \cap ((\bigcup_{i=1}^k (U_i \times \{y_i\})) \cup ((X \setminus A) \times \{y_0\})))),$$

where $E \subset X$ is a Borel set, and pr_1 denotes the projection onto the first factor. Direct calculations show that π is a coupling of μ and ν (we leave them to the reader). Moreover,

$$d_W(\mu, \nu) = \int_{X \times X} d(x, y)d\pi(x, y) \leq$$

$$\sum_{i=1}^k \text{diam}(U_i)\pi(U_i \times \{y_i\}) + \text{diam}(X)\pi(U_i \times \{y_i\}) \leq$$

$$\sum_{i=1}^k \mu(U_i)\varepsilon + \text{diam}(X)\mu(X \setminus A) \leq (1 + \text{diam}(X))\varepsilon.$$

This ends the proof. $\qquad\square$

Let us denote by $B(x, r)$ the open ball of diameter r and center in x, and let $N(B, \eta) = \bigcup_{x \in A} B(x, \eta)$.

Lemma 1.5 *Suppose that $d_W(\mu, \nu) < \varepsilon$. Then for any measurable set $A \subset X$*

$$\mu(A) \leq \nu(N(A, \sqrt{\varepsilon})) + \sqrt{\varepsilon} \quad \text{and} \quad \nu(A) \leq \mu(N(A, \sqrt{\varepsilon})) + \sqrt{\varepsilon}.$$

Proof Denote by $A_\eta = \{(x, y) \in X \times X : d(x, y) \geq \eta\}$. Observe that there exists a coupling $\pi \in \Pi(\mu, \nu)$ for which $\pi(A_{\sqrt{\varepsilon}}) < \sqrt{\varepsilon}$. Indeed, suppose conversely that $\pi(A_{\sqrt{\varepsilon}}) \geq \sqrt{\varepsilon}$ for all couplings $\pi \in \Pi(\mu, \nu)$. Since $d_W(\mu, \nu) = \inf_{\pi \in \Pi(\mu, \nu)}$, then for any coupling π we have

$$\int_{X \times X} d(x, y) d\pi(x, y) = \int_{X \setminus A_{\sqrt{\varepsilon}}} d(x, y) d\pi(x, y) + \int_{A_{\sqrt{\varepsilon}}} d(x, y) d\pi(x, y)$$

$$\geq \int_{A_{\sqrt{\varepsilon}}} d(x, y) d\pi(x, y) \geq \sqrt{\varepsilon} \cdot \pi(A_{\sqrt{\varepsilon}}) \geq \varepsilon.$$

This contradicts with the assumption $d_W(\mu, \nu) < \varepsilon$.

Next, let π be a coupling satisfying $\pi(A_{\sqrt{\varepsilon}}) < \sqrt{\varepsilon}$, and let $A \subset X$.

$$\mu(A) = \pi(A \times X) = \pi((A \times X) \cap A_{\sqrt{\varepsilon}}) + \pi((A \times X) \setminus A_{\sqrt{\varepsilon}})$$

$$\leq \pi((A \times X) \setminus A_{\sqrt{\varepsilon}}) + \sqrt{\varepsilon} = \pi(\{(x, y) \in X \times X : x \in A \text{ and } d(x, y) \leq \sqrt{\varepsilon}\} + \sqrt{\varepsilon}$$

$$\leq \pi(\{(x, y) \in X \times X : x \in A \text{ and } d(x, y) < \varepsilon\})$$

$$+ \pi(\{(x, y) \in X \times X : x \notin A \text{ and } d(x, y) < \varepsilon\}) + \sqrt{\varepsilon}$$

$$= \pi(X \times N(A, \sqrt{\varepsilon})) + \sqrt{\varepsilon} = \nu(N(A, \sqrt{\varepsilon}) + \sqrt{\varepsilon}.$$

The analogical calculation (with μ and ν reversed) gives $\nu(A) = \mu(N(A, \sqrt{\varepsilon}) + \sqrt{\varepsilon}$. This ends our proof.   □

Chapter 2
Foliations and Heat Diffusion

In this chapter we briefly recall the theory of foliations. We present the definition of a foliation and illustrate it by a number of examples (product foliation, a foliation given by a submersion, Reeb foliation of a solid torus, a linear foliation of a torus). We also recall the notion of the holonomy of a leaf. We present the foliated Laplace operator and foliated heat diffusion operators semigroup which plays important role in the metric diffusion. We only demonstrate these facts which are necessary for further results. For the complete theory one can refer to [3], one of the best books about foliations. We also take advantage of some results from [13, 17], and [21].

2.1 Basic Facts

A *foliation* $\mathscr{F}$ on an n-dimensional manifold is an equivalence relation with the equivalence classes being connected, immersed submanifolds, all of dimension $p \geq 1$. Locally, the decomposition into equivalence classes can be modelled on the decomposition of $\mathbb{R}^n$ into the cosets $x + \mathbb{R}^p$ of the imbedded subspace $\mathbb{R}^p$. The equivalence classes of $\mathscr{F}$ are called the *leaves* of $\mathscr{F}$.

More formally, let M be an arbitrary n-dimensional C^r-manifold. A C^r-*foliated chart* of codimension q is a pair (U, φ), where $\varphi : U \to V \times W$ is a C^r-diffeomorphism of an open subset $U \subset M$ into a set $V \times W \subset \mathbb{R}^{p+q}$, $p = n-q$, and V and W are rectangular neighborhoods in $\mathbb{R}^p$ and $\mathbb{R}^q$, respectively. The set $P_y = \varphi^{-1}(V \times \{y\})$, $y \in W$, is called a *plaque* of this chart.

Let $\mathscr{F} = \{L_\alpha\}_{\alpha \in \mathscr{A}}$ be a decomposition of M into connected, p-dimensional, $p = n - q$, topologically immersed submanifolds (*leaves*). Suppose that M admits an atlas $\{U_\lambda\}_{\lambda \in \Lambda}$ of foliated charts (Figure 2.1) of codimension q such that for each $\lambda \in \Lambda$ and $\alpha \in \mathscr{A}$ the set $L_\alpha \cap U_\lambda$ is a union of plaques. $\mathscr{F}$ is said to be *a foliation* of M, the number $p = n - q$ is called the *dimension* of $\mathscr{F}$, while q the *codimension*. A leaf passing through a point $x \in M$ will be later denoted by L_x.

© The Author(s) 2017
S.M. Walczak, *Metric Diffusion Along Foliations*, SpringerBriefs in Mathematics,
DOI 10.1007/978-3-319-57517-9_2

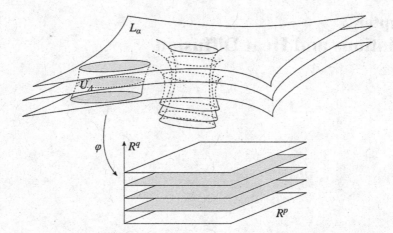

Fig. 2.1 A foliated chart

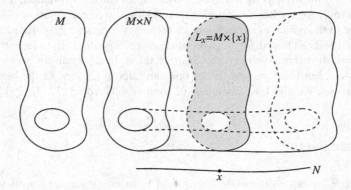

Fig. 2.2 The product foliation

If a foliated atlas is of class C^r, with $0 \le r \le \infty$ or $r = \omega$, then $\mathscr{F}$ and the foliated manifold $(M, \mathscr{F})$ are said to be of class C^r.

Example 2.1 The simplest example of a foliation is the *product foliation*. Let M and N be two arbitrary manifolds, $\dim M = m$ and $\dim N = n$. We foliate the product $M \times N$ by submanifolds $M \times \{x\}$, $x \in N$ (Figure 2.2).

Example 2.2 Let M and N be two arbitrary manifolds of dimension m and n, $m > n$, respectively. A smooth submersion $\pi : M \to N$ provides a foliation on M by the connected components of the non-empty level sets $\pi^{-1}(x)$, $x \in N$ (Figure 2.3).

Example 2.3 As an example of a foliation which is not provided by a single submersion, one can consider the *Reeb foliation* of a solid torus.

Fig. 2.3 A foliation defined by a submersion

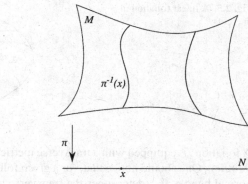

Fig. 2.4 Reeb foliation of a solid torus

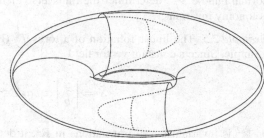

Let $M = D^2 \times S^1$, with D^2 denoting a two-dimensional disk, and S^1 being the unit circle. Let $f : (-1, 1) \to \mathbb{R}$ be a smooth function satisfying

$$f(0) = 0, \quad f(t) \geq 0, \quad f(t) = f(-t),$$

$$\lim_{t \to \pm 1} \frac{d^k}{dt^k} f(t) = \infty, \quad \lim_{t \to \pm 1} \frac{d^k}{dt^k} \left(\frac{1}{\frac{d}{dt} f(t)} \right) = 0,$$

for all $k = 0, 1, 2, \ldots$, for example $f(t) = e^{\frac{t^2}{1-t^2}} - 1$. For $\alpha \in [0, 1)$, define leaves of a foliation by

$$L_\alpha = \{(x, y) \in \text{int} D^2 \times S^1 : y = e^{2\pi(\alpha + f(\|x\|))i}\}.$$

Adding the boundary torus $\partial(D^2 \times S^1)$ as a boundary leaf we obtain a foliation of the solid torus $D^2 \times S^1$ called the *Reeb foliation* (Figure 2.4).

Example 2.4 ([13, Section 2.2]) Consider a foliation $\mathcal{F}$ of the codimension q on a manifold M, and denote by $\mathcal{L}$ the Lie derivative on M. A *transverse metric* on $(M, \mathcal{F})$ is a positive smooth bilinear form g on a module $\mathcal{X}(M)$ of all vector fields on M satisfying

1. $\text{Ker}(g_x) = T_x(\mathcal{F})$,
2. $L_X g = 0$ for any vector field X tangent to $\mathcal{F}$.

Fig. 2.5 A linear foliation of
the torus T^2

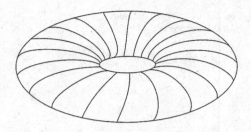

A foliation $\mathscr{F}$ equipped with a transverse metric g on $(M, \mathscr{F})$ is called a *Riemannian foliation*. One should notice that, for a given foliation, a Riemannian structure on the normal bundle $\mathscr{F}^\perp$ determines the transverse metric if and only if this structure is holonomy invariant.

Example 2.5 The linear foliation of a torus T^2 (Figure 2.5) is also an interesting example. Since a constant vector field

$$\tilde{X} = \begin{bmatrix} a \\ b \end{bmatrix}, \quad a, b \in \mathbb{R}$$

on $\mathbb{R}^2$ is invariant by all translations in $\mathbb{R}^2$, it defines a vector field X on $\mathbb{R}^2/\mathbb{Z}^2$. Assume that $a \neq 0$. The foliation $\tilde{\mathscr{F}}$ on $\mathbb{R}^2$ defined by integral curves (parallel lines of slope $\frac{b}{a}$) of $\tilde{X}$ passes to the foliation $\mathscr{F}$ on T^2 defined by X. Observe that for $\frac{b}{a}$ rational, $\mathscr{F}$ is a foliation of T^2 by circles. Otherwise, each leaf is a one-to-one immersion of $\mathbb{R}$ and is everywhere dense in T^2.

Example 2.6 *Compact foliations*, i.e., foliations with all leaves compact, is a family of foliations of our special interest. The topological structure of such foliations was deeply studied in [7] and [9]. Compact foliations will be the objects of deeper studies in Chapter 3.

2.2 Holonomy

Some important properties of the leaves of a foliation are described in terms of the *holonomy group* of a leaf, which describes the behavior of leafs in a small neighborhood of this leaf.. To understand the notion of holonomy group, we first recall the notion of a germ of a map.

Let M and N be manifolds, and let $x \in M$, $y \in N$. Consider a map $f : U \to V$, where U and V are open neighborhoods of x and y, respectively. A map $f' : U' \to V'$, $x \in U'$, $y \in V'$ is said to be equivalent to f iff there exists an open neighborhood $W \subset U \cap U'$ of x such that $f|W = f'|W$. The equivalence above defines the equivalence relation in the set of all mappings such that x is mapped to y. An equivalence class $[f]_x$ of this relation is called the *germ* of f at x.

Let us now consider mappings $f, g : M \to M$ preserving x, i.e., $f(x) = x$. One can compose the germs $[f]_x$ and $[g]_x$ into a germ $[f]_x \circ [g]_x = [f \circ g]_x$. The identity map defines the identity germ $[\mathrm{id}]_x$, and if f is a diffeomorphism on a neighborhood U of x, one can define an inverse germ $[f]_x^{-1} = [f^{-1}]_x$. Hence, the set of all germs of diffeomorphisms preserving x is a group, which will be denoted by Diff_x.

We now return to the definition of the holonomy group of a leaf. Let $L \in \mathscr{F}$ and $x, y \in L$. Consider a curve $\gamma : [0, 1] \to L$ linking x and y, that is a curve that $\gamma(0) = x$ and $\gamma(1) = y$. Let T_x and T_y be two transversals (smoothly imbedded, compact, connected, q-dimensional manifolds without boundary which are everywhere transverse to $\mathscr{F}$) through x and y, respectively. We associate with γ a germ at x of a diffeomorphism $[h]_{yx}(\gamma)$ defined as follows:

First, let us suppose that $\gamma([0, 1])$ is totally contained in a foliated chart. Since γ is a curve on L, x and y lie on the same plaque (we will denote by P_z the unique plaque containing $z \in M$). There exist an open neighborhood $A_x \subset T_x$ of x and a smooth map $h : A_x \to T_y$ with $h(x) = y$ and assigning for any $x' \in A$ the unique point $P_{x'} \cap T_y$. Moreover, A can be chosen in such a way that h is a diffeomorphism onto its image (Figure 2.6). We define

$$[h]_{yx}(\gamma) = [h]_x.$$

In the general case (Figure 2.7), let us choose a sequence of foliated charts $(U_1, \ldots, U_k)$ covering $\gamma([0, 1])$ for which $U_i \cap U_{i+1} \neq \emptyset$, and $\gamma([\frac{i-1}{k}, \frac{i}{k}]) \subset U_i$. Set $\gamma_i = \gamma|[\frac{i-1}{k}, \frac{i}{k}]$, $i = 1, \ldots, k$, and $x_i = \gamma(\frac{i-1}{k})$, $i = 0, \ldots, k$. We choose transversals T_i at x_i and define

$$[h]_{yx}(\gamma) = [h]_{x_k x_{k-1}}(\gamma_k) \circ \cdots \circ [h]_{x_1 x_0}(\gamma_1).$$

One can check [3] that the holonomy map h does not depend on a choice of a sequence $(U_1, \ldots, U_k)$. Moreover, if T_x, T_y, and T_z are transversals at $x, y, z \in L$,

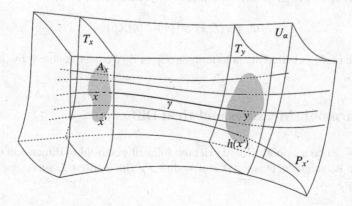

Fig. 2.6 Holonomy map in the single chart U_α

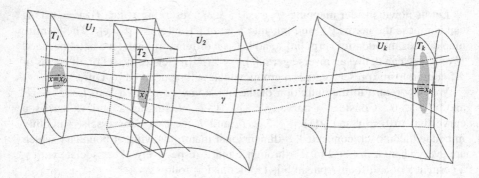

Fig. 2.7 The holonomy of a leaf

respectively, and $\gamma : [0, 1] \to L$ links y with z, while $\delta : [0, 1] \to L$ links x with y, then

$$[h]_{zx}(\gamma * \delta) = [h]_{zy}(\gamma) \circ [h]_{yx}(\delta),$$

where

$$(\gamma * \delta)(t) = \begin{cases} \gamma(2t) & \text{for } t \in [0, \frac{1}{2}), \\ \delta(2t - 1) & \text{for } t \in [\frac{1}{2}, 1]. \end{cases}$$

In addition, for any homotopic leaf curves γ and δ linking x with y, the corresponding holonomies are equivalent, i.e.,

$$[h]_{yx}(\gamma) = [h]_{yx}(\delta).$$

Let $\gamma : [0, 1] \to L$ be a loop at x, i.e., a curve with $\gamma(0) = \gamma(1) = x$. One can now define a holonomy homomorphism from the first fundamental group $\pi_1(L, x)$ at x into the group of germs of diffeomorphisms of a transversal T_x by

$$\Phi_L : \pi(L, x) \ni [\gamma] \mapsto [h]_x(\gamma).$$

The image of Φ_L is called the holonomy group of L. We later denote it by $\mathcal{H}_L$.

2.3 Harmonic Measures and Heat Diffusion

Let $(M, \mathcal{F}, g)$ be a smooth closed oriented foliated manifold of dimension n equipped with a Riemannian tensor g. Let $p = \dim \mathcal{F}$. The *leafwise Laplace operator* Δ defined by

$$\Delta f = \text{div} \nabla f$$

with ∇ denoting the gradient of f, has, in a foliated chart $U = D \times Z$ with coordinates $(x, z) = (x_1, \ldots, x_p, z)$ and leafwise metric tensor

$$g = \sum_{i,j=1}^{p} g_{i,j}(x_1, \ldots, x_p, z) \mathrm{d}x^i \otimes \mathrm{d}x^j,$$

a local expression

$$\Delta f = \frac{1}{\sqrt{|g|}} \sum_{j=1}^{p} \frac{\partial}{\partial x^j} \left(\sum_{i=1}^{p} g^{ij} \sqrt{|g|} \frac{\partial}{\partial x^i} f \right),$$

where (g^{ij}) denotes the inverse matrix of the Riemannian tensor matrix (g_{ij}) and $|g| = \det(g_{ij})$. Thus

$$\Delta = \sum_{i,j=1}^{p} g^{ij} \frac{\partial^2}{\partial x_i \partial x_j} + \text{first order terms.}$$

Since the metric tensor g on $(M, \mathcal{F})$ induces a metric tensor $g|_L$ on each leaf $L \in \mathcal{F}$, a leafwise Laplacian can be defined (following [3] or [21]) as

$$\Delta_L = \Delta|_L.$$

Define the *foliated Laplace operator* $\Delta_{\mathcal{F}}$ by

$$\Delta_{\mathcal{F}} f(x) = \Delta_{L_x} f(x), \quad x \in M,$$

where L_x is a leaf through x, and Δ_{L_x} is the Laplace operator on $(L_x, g|L_x)$. The operator $\Delta_{\mathcal{F}}$ acts on bounded measurable functions, which are C^2-smooth along the leaves.

We say that a probability measure μ on $(M, \mathcal{F}, g)$ is *harmonic* if

$$\int_M \Delta_{\mathcal{F}}(x) f d\mu(x) = 0 \text{ for any } f : M \to \mathbb{R}.$$

L. Garnet proved [11] that the harmonic measures are related to the differential operator $\Delta_{\mathcal{F}}$. First, we formulate the following.

Theorem 2.1 *On any compact foliated Riemannian manifold, harmonic probability measure exists.*

Lemma 2.1 *Let (X, g) be a Riemannian manifold, and let Δ denote the Laplace operator. If f is a function on M of class C^2, and $x_0 \in M$ is a local maximum of f, then $\Delta f(x_0) \leq 0$. If $x_0 \in M$ is a local minimum of f, then $\Delta f(x_0) \geq 0$.*

Proof Let x_0 be a local maximum of f. Since

$$\Delta f = \sum_{i,j} g^{ij} \frac{\partial^2 f}{\partial x_i \partial x_j} + \text{first order terms},$$

we can choose the coordinate system that $g^{ij}(x_0)$ is the identity matrix. Moreover, since Δ annihilates constants and all first order derivatives vanish at x_0, then

$$\Delta f(x_0) = \sum_{i,j} g^{ij} \frac{\partial^2 f}{\partial x_i \partial x_j}(x_0) \leq 0.$$

The second assertion goes in the same way.   □

Let $C(M)$ denote the space of continuous functions on M, while $\mathbf{1}$ a constant function $\mathbf{1}(x) = 1$, for all $x \in M$.

Lemma 2.2 *On a compact foliated Riemannian manifold M, the closure of the range of $\Delta_{\mathscr{F}}$ does not contain $\mathbf{1}$.*

Proof Suppose that there is a sequence $\{f_i\}_{i \in \mathbb{N}}$ such that $\Delta_{\mathscr{F}} f_i$ converges in $C(M)$ to $\mathbf{1}$. So, there exists an index $i_0 \in \mathbb{N}$ such that for all $i > i_0$

$$\Delta_{\mathscr{F}}(x) f_i \geq \frac{1}{2} \quad \text{for all } x \in M.$$

Since M is compact, by Lemma 2.1, $\Delta_{\mathscr{F}}(x) f_i \leq 0$ somewhere in M.   □

Lemma 2.3 *A continuous linear functional $\Phi : C(M) \to \mathbb{R}$ is given by the integral with respect to the probability measure μ on M if and only if $\|\Phi\| = 1$ and $\Phi(\mathbf{1}) = 1$.*

Proof To begin, assume that $\|\Phi\| = 1$ and $\Phi(\mathbf{1}) = 1$. By the Riesz Representation Theorem [15], it is only necessary to show that for nonnegative functions $\Phi(f) \geq 0$. We can assume that $0 \leq f \leq 1$, for all $x \in M$. The function $h = 2f - \mathbf{1}$ satisfies $-\mathbf{1} \leq h \leq \mathbf{1}$. Hence $|\Phi(h)| \leq 1$, and $\Phi(h) \geq -1$. This implies $2\Phi(f) \geq 0$.

The converse is immediate.   □

Proof (Theorem 2.1) Let $H \subset C(M)$ be the closure of the range of $\Delta_{\mathscr{F}}$ in the uniform norm. Let $a = \inf_{f \in H} \|\mathbf{1} - f\|$. By Lemma 2.2, $a > 0$. Moreover, $a \leq 1$ because H is a subspace, and $a \geq 1$ due to the fact that for any continuous function which is C^2 along the leaves $\mathbf{1} - \Delta g \leq 1$ somewhere on M (due to Lemma 2.1).

Let $\Phi : H + \mathbb{R}\mathbf{1} \to \mathbb{R}$ be the linear functional defined by $\Phi(h + t\mathbf{1}) = t$. For $v = h + t\mathbf{1} \in H + \mathbb{R}\mathbf{1}$ and $t = 0$ we have $|\Phi(v)| = 0 \leq \|v\|$. If $t \neq 0$, then

$$|\Phi(v)| = |t| \leq |t|(\|\frac{1}{t}h + \mathbf{1}\|) = \|v\|.$$

So, by the Hahn–Banach Theorem, there exists a linear extension $\Psi : C(M) \to \mathbb{R}$ of Φ, such that $|\Psi(g)| \leq \|g\|$ for all $g \in C(M)$. Moreover, $\Psi(1) = 1$, so $\|\Psi\| = 1$. In addition,

$$\Psi|_H = \Phi_H \equiv 0.$$

By Lemma 2.3, Ψ is the integral associated to the probability measure μ on M, which is the desired harmonic measure. □

Let f be a bounded continuous function on a manifold L. Recall that if the geometry of L is bounded, one can solve on L the heat equation

$$\frac{\partial}{\partial t}u(x,t) = \Delta u(x,t)$$

with the initial condition f, where $u \in C^{2,1}(L \times [0,\infty))$, and $u(x,0) = f(x)$.

Let L be a leaf of $\mathscr{F}$. The heat equation on $(L, g|_L)$ admits a fundamental solution $p_t(x,y)$, called the *heat kernel*, which satisfies

$$\frac{\partial}{\partial t}p_t(x,y) = \Delta_x p_t(x,y) \text{ for any } y \in L,$$

and for any bounded function f on L

$$D_{L,t}f(x) = \int_L f(y)p_t(x,y)dy$$

is the bounded solution to the heat equation on L with the initial condition f. The operators $D_{L,t}$ form the *semigroup of diffusion operators* on $(L, g|L)$. The aggregate of $D_{L,t}$ on various leaves defines on M a semigroup D_t of operators satisfying on functions on M

$$D_0 = \text{id}, \ D_{t+s} = D_t \circ D_s, \ \frac{d}{dt}D_t|_{t=0} = \Delta_{\mathscr{F}}.$$

Each D_t restricted to a leaf $L \in \mathscr{F}$ coincides with the heat diffusion operators on L. Thus, for suitable functions f on M, $D_t f$ is a function defined at any $x \in M$ by

$$(D_t f)(x) = \int_{L_x} f(y)p_t(x,y)dy$$

with $p_t(x,y)$ being the heat kernel on $(L_x, g|_{L_x})$.

Let μ be a probability measure on M. Following [3] and [21], one can define *the diffused measure* $D_t\mu$ by the formula

$$\int_M f(x)dD_t\mu(x) = \int_M D_t f(x)d\mu(x),$$

where f is bounded measurable function on M. A measure μ is called *diffusion invariant* when $D_t\mu = \mu$.

In addition, we present the important properties of the heat kernel on real line and circle that will be later needed.

Remark 2.1 Let $p_t(x, y)$ be the heat kernel on $\mathbb{R}$, that is

$$p_t(x, y) = \frac{1}{\sqrt{4\pi t}} e^{-\frac{|x-y|^2}{4t}}.$$

Let $R > 0$, $R' > R$ and $t > 0$. One can check that for $s \geq 0$ satisfying $R^2 s = (R')^2 t$

$$\int_{kR}^{(k+1)R} p_t(x, 0)dx = \int_{kR'}^{(k+1)R'} p_s(x, 0)dx$$

for all $k \in \mathbb{Z}$. Let $\epsilon > 0$. Suppose that $R = l\xi\epsilon$, and $R' = l'\xi\epsilon$, $l, l', \xi \in \mathbb{N}$. Define sets

$$A_i^k = \bigcup_{j=0}^{l-1} [kR + (j\xi + i)\epsilon, kR + (il + j + 1)\epsilon]$$

$$B_i^k = \bigcup_{j=0}^{l'-1} [kR + (jl\xi + i)\epsilon, kR + (il + j + 1)\epsilon]$$

where $k \in \mathbb{Z}$ and $i = 0, 1, \ldots, \xi - 1$. Due to uniform equicontinuity of the heat kernel $p_t(x, 0)$ on $\mathbb{R}$, one finds that there exists $T > 0$ such that for all $s, t > T$ satisfying $R^2 s = (R')^2 t$

$$\sum_{k\in\mathbb{Z}} \sum_{i=0}^{\xi-1} |\int_{A_i^k} p_t(x, 0)dx - \int_{B_i^k} p_s(x, 0)dx| \leq \epsilon.$$

Remark 2.2 Due to the form of the heat kernel on circle, which is

$$P_t(x, y) = \sum_{n\in\mathbb{Z}} p_t(x + rn, y),$$

with $p_t(x, y)$ being the heat kernel in $\mathbb{R}$, we can extend Remark 2.1 to the sets of the form A_i^k and B_i^k, $k = 0, \ldots, m-1$, $i = 1, \ldots, \xi - 1$, on circles of the length mR and mR', respectively.

For a detailed description of the theory of harmonic measures and foliated heat diffusion, one should refer to [2], where it is described in full generality in terms of foliated spaces.

Chapter 3
Compact Foliations

Compact foliations on compact manifolds, i.e., foliations with all leaves compact, from our point of view, are very interesting. In 1952, Reeb [14] described a smooth flow on a non-compact manifold which has periodic orbits such that the length of orbits is locally unbounded. Edwards, Millet, and Sullivan in [7] have given a full answer to the question (named the Periodic Orbit Conjecture) on the existence of an upper bound on the volume of the leaves on a compact manifold M foliated by compact submanifolds.

In general, the answer is negative, as the following section shows. First, we present two examples of foliations by circles of compact manifolds for which the length of the leaves is unbounded. In this chapter, we recall some important facts about the relation between the topology of the leaf space of compact foliations and the volume function which have the great influence on metric diffusion.

3.1 Examples

The first example of a foliation with the volume of leaves unbounded is the one in codimension 4, which was presented by D. Sullivan in [16]. The second is the example in the lowest possible codimension, that is in the codimension 3. It was published by D.B.A. Epstein and E. Vogt [10] in 1978. We here present only the precise analytic description of the manifold and foliation given in [10], Sections 3, 4, and 6.

Example 3.1 Consider deformations (through immersed curves) of the 2-dimensional sphere S^2 as shown in Figure 3.1. Repeating the operations one can produce a moving curve γ_t, $t \in [1, \infty)$, on S^2 satisfying:

1. the geodesic curvature of γ_t go uniformly to infinity as $t \to \infty$,
2. for any $t \in [1, 2)$ the Hausdorff limit of $(\gamma_{t+n})_{n \in \mathbb{N}}$ is γ_t,

© The Author(s) 2017
S.M. Walczak, *Metric Diffusion Along Foliations*, SpringerBriefs in Mathematics,
DOI 10.1007/978-3-319-57517-9_3

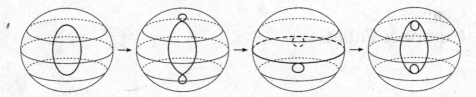

Fig. 3.1 Twist of a curve

Fig. 3.2 The clocks

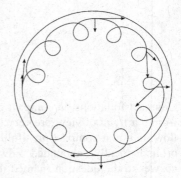

3. for any $t \in [1, 2)$ a sequence $(l(\gamma_{t+n}))_{n \in \mathbb{N}}$, where $l(\gamma)$ denotes the length of a curve γ, is an increasing sequence satisfying

$$\frac{1}{n} \leq l(\gamma_{t+n+1}) - l(\gamma_{t+n}) \leq \frac{2}{n}.$$

For fixed time $t \geq 1$, we consider all congruent curves γ_t^α of γ_t obtained by rotations of S^2 by elements $\alpha \in SO_3$. We add to γ_t a vector field of constant length $\frac{1}{t}$ uniformly turning exactly once around the tangent vector as a point traces the curve with constant speed. Adding the unit tangent vector field we obtain, after rotating by elements of SO_3, a four-dimensional family of *curves of clocks* on S^2 (Figure 3.2).

Let

$$M_t = \{(x, v_1, v_2) : x \in S^2, v_1, v_2 \in T_x S^2 \text{ and } |v_1| = 1, |v_2| = \frac{1}{t}\}.$$

For all $t \in [1, \infty)$, M_t fill up the deleted neighborhood of the unit tangent bundle $T = T^1 S^2$ of the two-dimensional sphere. Thus,

$$M = T \cup \bigcup_{t \in [1, \infty)} M_t$$

is in a natural way a compact five-dimensional Riemannian manifold with boundary M_1, and with M_t at a uniform distance of $\frac{1}{t}$ from T.

The clock structure on γ_t and its images under rotations by the elements α of SO_3 give a three-dimensional family of curves γ_t^α in M_t, $t \in [1, \infty)$, $\alpha \in SO_3$. The fibers of the natural projection $T \to S_2$ define a family of circles on T. γ_t^α with fixed t and α varying in SO_3 exactly fill M_t by embedded circles since each clock (x, v_1, v_2) appears exactly ones in γ_t^α, and define a compact foliation on M. The length of γ_{t+n}^α goes to infinity as $n \to \infty$, while the tangent directions of γ_{t+n}^α approach the tangent directions of circles on T. In addition, there exists a constant $C > 0$ for which

$$l(\gamma_{t+n+1}^\alpha) - l(\gamma_{t+n}^\alpha) < C$$

for all $t \in [1, 2)$ and $\alpha \in SO_3$.

Example 3.2 Let us consider the octagon

$$D = \{(x, y) \in \mathbb{R}^2 : |x| \le 2, |y| \le 2, |x - y| \le 3, |x + y| \le 3\},$$

and let

$$\psi(x, y) = (2 - x)(2 + x)(2 - y)(2 + y)(3 + x + y)(3 - x - y)(3 + x - y)(3 - x + y).$$

Notice that $\psi > 0$ on the interior of D, and $\psi|_{\partial D} = 0$. Let

$$A = D \cap \{(x, y) \in \mathbb{R} : \psi(x, y) \le 1\},$$

and let $\xi = (x, y, u_1, u_2, w_1, w_2, z) \in \mathbb{R}^7$. We define $F : \mathbb{R}^7 \to \mathbb{R}^3$ by

$$F_1(\xi) = u_1^2 + u_2^2 - 4 + x^2,$$

$$F_2(\xi) = w_1^2 + w_2^2 - 4 + y^2,$$

$$F_3(\xi) = z^2 - \varrho(x, y),$$

where

$$\varrho(x, y) = (1 - \psi(x, y))(3 - x - y)(3 + x + y)(3 + x - y)(3 - x + y).$$

Let $M = F^{-1}(0)$. The projection of $\mathbb{R}^7$ onto the first two coordinates maps the manifold M onto A. Thus M is a four-dimensional compact manifold. Define a vector field X on $\mathbb{R}^7$ by

$$X_\xi = \psi \frac{\partial \psi}{\partial y} \frac{\partial}{\partial x} - \psi \frac{\partial \psi}{\partial x} \frac{\partial}{\partial y}$$

$$+ (Ku_1 - pu_2) \frac{\partial}{\partial u_1} + (pu_1 + Ku_2) \frac{\partial}{\partial u_2}$$

$$+ (Lw_1 - qw_2) \frac{\partial}{\partial w_1} + (qw_1 + Lw_2) \frac{\partial}{\partial u_2} + z\sigma \frac{\partial}{\partial z},$$

where

$$K(x,y) = -x\frac{\partial\psi}{\partial y}(4 - y^2)(9 - (x + y)^2)(9 - (x - y)^2),$$

$$L(x,y) = y\frac{\partial\psi}{\partial y}(4 - x^2)(9 - (x + y)^2)(9 - (x - y)^2),$$

$$p(x,y) = (9 + x^2 - y^2)y,$$

$$q(x,y) = (9 - x^2 + y^2)x,$$

$$\sigma(x,y) = \left(\frac{\partial\psi(x,y)}{\partial x} - \frac{\partial\psi(x,y)}{\partial y}\right)\left(\frac{(x + y)\psi}{9 - (x + y)^2}\right)$$
$$+ \left(\frac{\partial\psi}{\partial x} + \frac{\partial\psi(x,y)}{\partial y}\right)\left(\frac{(y - x)\psi(x,y)}{9 - (x - y)^2}\right).$$

If $\xi \in M$, then $X(\xi) \in T_\xi M$. Hence, X is a nowhere vanishing vector field on M. Moreover, the orbit of X through ξ is diffeomorphic to circle if only $\psi(x,y) > 0$, the length of the orbit of the point ξ tends to infinity if $\psi(x,y)$ tends to zero, and all other orbits are also diffeomorphic to circle.

The existence of an upper bound on the volume of leaves has important consequences both on local and global topological structure of a foliation. This will be the subject of our interest in the next section, and will have a great influence on the metric diffusion.

3.2   Topology of the Leaf Space

Let $\mathscr{F}$ be a compact foliation, i.e., a foliation with all leaves compact, on a given compact manifold of dimension n. The space of all equivalence classes of the relation

$$x \sim y \Leftrightarrow L_x = L_y,$$

with quotient topology, is called *the leaf space*, and will be denoted by $\mathscr{L}$. Following [9], define the volume function $v : \mathscr{L} \to [0, \infty)$ according to the following procedure:

Let $\{U_i\}_{i\in\mathbb{I}}$ be a locally finite open covering of M, and let $\{\varphi_i : V_i \times W_i \to U_i\}_{i\in\mathbb{I}}$ be a family of foliated charts. For every $i \in \mathbb{I}$ let $v_i : W_i \to [0, \infty)$ be a continuous function. Suppose that the set $W_i' = \{w \in W_i : v_i(w) > 0\}$ has compact closure in W_i and the family of open sets $\varphi_i(V_i \times W_i')$ is a covering of M. Let $L \in \mathscr{F}$ be a leaf. For each $i \in \mathbb{I}$, define $L(i) = \mathrm{pr}_2(\varphi^{-1}(U_i \cap L)) \subset W_i$, where pr_2 denotes the projection onto the second factor. Finally, set

$$v(L) = \sum_{i \in \mathbb{I}} \sum_{w \in L(i)} v_i(w).$$

Denote by $\pi : M \to \mathscr{L}$ the quotient projection defined by $\pi(x) = L_x$. A subset A of M is called *saturated*, if it is the union of leaves of $\mathscr{F}$. The set $\hat{A} = \pi^{-1}(\pi(A))$ is called the *saturation* of A.

In order to study the topology of a compact foliation we present a few theorems about the connection between the volume function and the leaf space topology. Since the deep understanding of the topology of a leaf space is essential in further considerations, we recall the proofs as they explain the nature of compact foliations.

Theorem 3.1 *The following conditions are equivalent:*

1. π *is a closed map.*
2. π *maps compact sets onto closed sets.*
3. *Each leaf has arbitrarily small saturated neighborhoods.*
4. $\mathscr{L}$ *with quotient topology is Hausdorff.*
5. *If $K \subseteq M$ is compact, then the saturation of K is also compact.*

Proof The implication $(1) \Rightarrow (2)$ is obvious. Now, let us suppose that (2) is true. Let L be a leaf, U a compact neighborhood of L, and let K be a boundary of U. Then the saturation $\hat{K} = \pi^{-1}(\pi(K))$ is compact. Let $x \in \hat{U} \setminus \hat{K}$. Then $L_x \cap U \neq \emptyset$ and $L_x \cap K = \emptyset$. Since L_x is connected, it lies entirely in U. Hence $U \setminus \hat{K}$ is saturated, because it equals to $\hat{U} \setminus \hat{K}$, and open, since it equals to $\mathrm{int}(U \setminus \hat{K})$. This proves (3).

Obviously $(3) \Rightarrow (4)$. Now, suppose that (4) is true. Clearly, due to the previous part of this proof, π is a closed map. Since π is continuous, and the preimage of a single point is compact then π is proper.

Indeed, let $K \subset \mathscr{L}$ be a compact set, and let $\{U_\lambda\}_{\lambda \in \mathbb{L}}$ be an open covering of $\pi^{-1}(K)$. It is also a covering of $\pi^{-1}(y)$ for any $y \in K$. Since $\pi^{-1}(y)$ is compact, it has a finite sub-cover. Hence, for any $y \in K$, we can choose a finite subset $\lambda_y \in \mathbb{L}$ such that $\pi^{-1}(y) \subset \bigcup_{\lambda \in \lambda_y} U_\lambda$. The set $A_y = X \setminus \bigcup_{\lambda \in \lambda_y} U_\lambda$ is closed, and $\pi(A)$ is also closed in $\mathscr{L}$, since π is a closed map. Thus, the set $V_y = \mathscr{L} \setminus \pi(A_y)$ is open, and $y \in A_y$. Since K is compact, and $K \subset \bigcup_{y \in K} V_y$, we can choose a finite number of points $y_1, \dots, y_k \in K$ such that $K \subset \bigcup_{i=1}^{k} V_{y_i}$. Furthermore, $\Lambda = \bigcup_{i=1}^{k} \lambda_{y_i}$ is also finite,

$$\pi^{-1}(K) \subset \pi^{-1}\left(\bigcup_{i=1}^{k} V_{y_i}\right) \subset \bigcup_{\lambda \in \mathscr{L}} U_\lambda,$$

so $\pi^{-1}(K)$ is compact. This gives (5).

Finally, let (5) be true. For a closed set $A \subset M$ choose x in the closure of $\hat{A}$. By our hypothesis, there exists a compact saturated neighborhood K of x. Then the set

$B = K \cap A$ is compact, so $\hat{B}$ is also compact. We claim that $x \in \hat{B} \subseteq \hat{A}$. If not then $K \setminus \hat{B}$ would be saturated neighborhood of x disjoint from A and hence from $\hat{A}$. This completes the proof.   □

Recall that a Riemannian manifold is a pair (M, g) consisting of a differentiable manifold M and a smooth inner product g defined on the tangent bundle, which is called a Riemannian structure. If M is a foliated manifold, then g induces a Riemannian structure g_L on the tangent bundle of every leaf. Let (M, g) be a Riemannian manifold and let L be a submanifold on M. The Riemannian structure on M induces the volume of L (via the induced Riemannian structure g_L).

Let $X \subset M$ be a locally compact saturated set on a compact manifold carrying a compact foliation. Given $x \in X$, we will use a fixed tubular neighborhood W of L_x, and the bundle projection $\pi : W \to L_x$ containing the transverse q-disk $D_x = \pi^{-1}(x)$.

Theorem 3.2 *The following conditions are equivalent:*

1. *The restricted volume function* $\mathrm{vol}|_X$ *is bounded on some neighborhood N of any* $x \in X$.
2. *The restricted holonomy group* $\mathcal{H}_{L_x}$ *is finite.*
3. *There exists a transversal neighborhood V of $x \in D_x$ such that each holonomy mapping along any loop in L_x carries V onto itself, and the automorphism group* $\mathcal{H}_V$ *of V so produced is finite and isomorphic to the holonomy group* $\mathcal{H}_L$.

Proof $(2) \Rightarrow (3)$ follows from the general fact about germs. Let

$$\{f_i : V_i \to D_x \cap X\}_{i \in \{0,\dots,k\}}$$

be a finite collection of open embeddings of open neighborhoods V_i of x in $D_x \cap X$ with distinct germs $[f_i]$ such that they comprise the holonomy group $\mathcal{H}_{L_x}$. We assume that f_0 is the identity on its domain V_0. For each indices $i, j \in \{0, \dots, k\}$, denote by $k(i, j)$ the unique index such that $[f_i] \circ [f_j] = [f_k]$. By shrinking the domains V_i if necessary, we can assume that for each pair i, j

$$f_i f_j|_{V_i \cap f^{-1}(V_j) \cap V_k} = f_k|_{V_i \cap f^{-1}(V_j) \cap V_k}.$$

Define $V = \bigcap_{i,j}(V_i \cap f^{-1}(V_j))$ and $h_i = f_i|V$. Then $\mathcal{H}_V \cong \{h_i\}_{i \in \{0,\dots,n\}}$ is a group of automorphisms of V, isomorphic to the holonomy group $\mathcal{H}_{L_x}|_X$.

The implication $(3) \Rightarrow (2)$ follows directly from the definition of the holonomy group. In order to prove $(3) \Rightarrow (1)$, let U be the union of all leaves intersecting an open transversal V in X. Note that for sufficiently small V, we have $U \subset W$. Suppose that vol is not bounded on any neighborhood of L. Hence there must be a leaf F in U intersecting V in an arbitrarily large finite collection of points $\{y_0, \dots, y_n\}$. For each j, there is a path in F linking y_0 with y_j, hence there exists a holonomy map $h_j : V \to V$ such that $h(y_0) = y_j$. Finally, $\mathcal{H}_V$ must be arbitrarily large, and hence infinite. This contradicts (3).

To prove (1) $\Rightarrow$ (3) suppose that there is a bound on the volume $\mathrm{vol}(L_y)$ of the leaves passing through $y \in X$ near x. Then, given a neighborhood U_0 of L_x in $W \cap X$, there is a smaller neighborhood U_1 such that union U of all leaves in X which intersects U_1 lies in U_0. Otherwise, there would be leaves in X which intersect both $X \setminus U_0$ and any arbitrarily small U_i, which would force these leaves to have large volume. Now U is an open saturated neighborhood of L_x in U_0. Its intersection V with D_x gives a subset of $D_x \cap X$ which is mapped homomorphically onto itself by every holonomy mapping of L_x. Denote so produced group of automorphisms by $\mathscr{H}_V$.

By the assumption on bounded volume, we assume that V is so small that there is an integer n such that each leaf L_y, $y \in V$, intersects V in n or fewer points. Hence, any orbit of V under $\mathscr{H}_V$ has n or fewer points. Now, for each orbit $\Phi \subset V$, number its points $\Phi = \{y_1, \ldots, y_{n(\Phi)}\}$, where $n(\Phi) \leq n$. For each Φ, this provides a homomorphism $H_V \rightarrow S(n(\Phi)) \subset S(n)$ to the symmetry group $S(n)$, by restricting to Φ. There are only finitely many such homomorphisms, since H_V is finitely generated. Therefore we can group the orbits into finitely many disjoint collections $\{\Phi\}_i$, according to the homomorphism each orbit determines. Moreover, each homomorphism $h \in \mathscr{H}_V$ is determined by its images $\{h_i\}$ in $S(n)$, one image h_i for each collection $\{\Phi\}_i$, so h is determined by finite numbers if choices from the finite group $S(n)$. Hence $\mathscr{H}_V$ is finite. This ends the proof. $\qquad\square$

Theorem 3.3 *The restricted volume function*

$$\mathrm{vol}|_X : X \rightarrow (0, \infty)$$

is lower semi-continuous at any $x \in X$ in the following way:

For any integer $n > 0$, any $\epsilon > 0$, and any y in a sufficiently small neighborhood of x in X either

1. $\mathrm{vol}(L_y) > n \cdot \mathrm{vol}(L_x)$, or
2. *there exists an integer j, $1 \leq j \leq n$, such that*

$$|\mathrm{vol}(L_y) - j\mathrm{vol}(L_x)| < \epsilon.$$

Proof Observe that for y sufficiently close to x, the image $p(W \cap L_y)$ must cover all of L_x. In fact, fix an integer $n > 0$. Then if y is sufficiently close to x, it must be either

1. $p|_{W \cap L_y}$ is greater than n-to-1 everywhere or
2. $p|_{W \cap L_y}$ is a j-to-1 covering projection, for some $1 \leq j \leq n$.

This ends our proof. $\qquad\square$

Corollary 3.1 *A subset of X consisting of all points of continuity of $\mathrm{vol}|X$ is an open dense subset of X (Figure 3.3).*

Fig. 3.3 Graphical
presentation of the lower
semi-continuity of the volume
function (due to D.B.A.
Epstein [8])

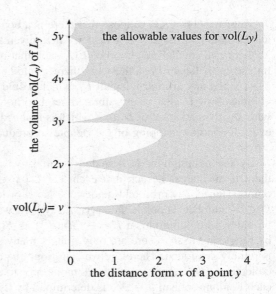

Proof The corollary follows directly from the fact that the set of all continuity points of a semi-continuous function on a locally compact space contains a dense subset, in general G_δ, but in our case open.   □

Theorem 3.4 *The following conditions are equivalent:*

1. *The restricted volume function* $\mathrm{vol}|_X$ *is continuous at* $x \in X$.
2. *The restricted holonomy group* $\mathscr{H}_{L_x}|_X$ *is trivial.*
3. *There exists a transversal neighborhood* V *of* x *in* $D_x \cap X$ *such that each holonomy map along any loop in* L_x *carries* V *identically onto itself.*

Proof Each of the above conditions is equivalent to the fact that for any y sufficiently close to x, the leaf L_y intersects D_x in exactly one point.   □

Remark 3.1 If the conditions of Theorem 3.2 hold, then the union of all leaves which intersect sufficiently small open transversal V provides an open saturated neighborhood U of L_x in $W \cap X$ such that the restriction of p to U is a foliated fiber bundle, with fiber V and group $\mathscr{H}_V$. This applies also for Theorem 3.4.

Remark 3.2 The conditions of Theorem 3.1, under the assumption that $\mathscr{F}$ is a C^1-foliation on a C^∞-manifold, provide the following local model for $\mathscr{F}$ near a leaf L:

Let q be a codimension. We are given a finite subgroup Γ of the orthogonal group $O(q)$ and a homomorphism $\psi : \pi_1(L) \to \Gamma$. If $\tilde{L}$ is a covering space of L corresponding to the kernel of Γ, then Γ acts on $\tilde{L}$ by covering translations. Let $\tilde{L} \times_\Gamma D^q$ be the quotient of $\tilde{L} \times D^q$ under the relation identifying $(\tilde{l}\gamma, p)$ with $(\tilde{l}, \gamma p)$ for each $\tilde{l} \in \tilde{L}$, $\gamma \in \Gamma$ and $p \in D^q$.

Let $\pi : \tilde{L} \times D^q \to \tilde{L} \times_\Gamma D^q$ be the quotient projection. Then $\tilde{L} \times_\Gamma D^q$ is foliated by compact sets of the form $\pi(\tilde{L} \times \{p\})$.

In other words, locally $\mathscr{F}$ is a foliated bundle with finite group.

Following Epstein [8] and Vogt [20], for a foliation with all leaves compact we can introduce the hierarchy of *bad sets*.

First, we fix a Riemannian metric on M and set $B_0 = M$. For any ordinal number $\alpha > 0$ we define

1. if α is a limit ordinal, we set $B_\alpha = \bigcap_{\beta < \alpha} B_\beta$,
2. for all non-limit ordinals B_α is the set of all $x \in B_{\alpha-1}$ such that the restricted volume $\mathrm{vol}|_{B_{\alpha-1}}$ is locally unbounded.

Note that B_β is nowhere dense in B_α, if $\beta > \alpha$. The sets B_α establish the natural filtration of the bad set

$$M = B_0 \supset B_1 \supset B_2 \supset \dots$$

which is called the *Epstein filtration* or the *Epstein hierarchy*. Every B_α is saturated and closed. Since M has a countable base and is compact, there is a countable ordinal γ such that $B_{\gamma+1} = \emptyset$ and $B_\gamma \neq \emptyset$. The ordinal γ is called the *length of the Epstein hierarchy* and will be later denoted by $\mathrm{Eps}(\mathscr{F})$.

By Remark 3.2, the structure of the set $G = B_0 \setminus B_1$, i.e., the set of all points $x \in M$ near which the volume function is bounded, is well known. All the leaves contained in G has finite holonomy, and $G/\mathscr{F}$ is Hausdorff. Moreover, it is open, saturated, and dense in M (by Corollary 3.1). The set G is called *the good set* of $\mathscr{F}$.

Chapter 4
Metric Diffusion

This chapter is devoted to the notion of the metric diffusion along foliations. The Wasserstein distance, theory of foliations, and foliated heat diffusion which we discussed in the previous chapters lie beneath this notion.

The idea of metric diffusion along a foliation was proposed by Jesus A. Alvarez-Lopez from the University of Santiago de Compostela, and comes from the earlier observations [23] of the author of this work that warped foliations, i.e., a foliated Riemannian manifold with a metric modified conformally along the leaves and left unchanged in the direction perpendicular to the leaves, can lose (in some sense) compact leaves. Precisely speaking, if $(f_n : M \to \mathbb{R})_{n \in \mathbb{N}}$ is a sequence of warping functions, that is a sequence of functions constant along the leaves, converging uniformly to zero, and $\mathscr{F}$ is a compact foliation with non-empty connected bad set B, then B may collapse in Gromov–Hausdorff sense, as n goes to infinity, to the singleton.

In this chapter, we propose a new family $\{D_t d\}_{t \geq 0}$ of metrics on a compact foliated Riemannian manifold as the Wasserstein distance of Dirac masses diffused at time $t > 0$ by the foliated diffusion of measures.

4.1 Metric Diffusion

Let $(M, \mathscr{F}, g)$ be a smooth compact foliated manifold equipped with a Riemannian metric g and carrying a foliation $\mathscr{F}$. Let δ_x denote the Dirac mass concentrated at x, and let $t > 0$ be a real number. We define

$$D_t d(x, y) := d_W(D_t \delta_x, D_t \delta_y),$$

with D_t being the foliated heat diffusion operator. $D_t d$ will be called *the metric diffused along $\mathscr{F}$*.

© The Author(s) 2017
S.M. Walczak, *Metric Diffusion Along Foliations*, SpringerBriefs in Mathematics,
DOI 10.1007/978-3-319-57517-9_4

Since $D_t \delta_x \neq D_t \delta_y$ if and only if $x \neq y$, and d_W is a metric on $\mathscr{P}(M)$ then $D_t d$ is a metric on M. Since $d_W(\delta_x, \delta_y) = d(x, y)$ for any two points $x, y \in M$ and $D_0 = \mathrm{id}$, $D_0 d$ coincides with the original metric d induced by the Riemannian structure g. The metric space $(M, D_t d)$, that is M equipped with metric d diffused along $\mathscr{F}$, will be later denoted by M_t.

Following [6], we recall that two metrics ρ_1 and ρ_2 are called *equivalent* if they induce the same topology. In other words, ρ_1 and ρ_2 are equivalent if and only if they induce the same convergence, i.e. for any $x \in X$ and any sequence $(x_i)_{i \in \mathbb{N}}$

$$\lim_{i \to \infty} \rho_1(x_i, x) = 0 \Leftrightarrow \lim_{i \to \infty} \rho_2(x_i, x) = 0.$$

Theorem 4.1 *For any $t \geq 0$, metrics $D_t d$ and $D_0 d$ are equivalent.*

Proof Let $t > 0$, $x \in M$, and let $(x_i)_{i \in \mathbb{N}}$ be a sequence D_0-converging to x, i.e., $\lim_{i \to \infty} d(x_i, x) = 0$. Let $1 > \varepsilon > 0$. Denote by $\mu_p = D_t \delta_p$. Let us choose a closed d-ball

$$B_{L_x}(x, R) = \{y \in L_x : d_{L_x}(x, y) < R\}$$

on the leaf L_x (d_{L_x} denotes the induced Riemannian metric on L_x) such that $\mu_x(B_{L_x}(x, R)) > 1 - \varepsilon$. Consider a foliated tubular neighborhood $N(B_{L_x}(x, R), \varepsilon)$ of $B_{L_x}(x, R)$ with the diameter smaller than ε. Choose a partition $\{U_0, \ldots, U_k\}$ of $B_{L_x}(x, R)$ by the pairwise disjoint subsets with $x \in U_0$ satisfying $\mathrm{diam}U_j < \varepsilon$ and $\mu_x(U_j) \leq \varepsilon$ for every $0 \leq j \leq k$. Next, for every $j = 1, \ldots, k$ we choose $y_j \in U_j$, we set $y_0 = x$, and we lift the partition by U_j's to the partition $\{V_0, \ldots, V_k\}$ of $N((B_{L_x}(x, R), \varepsilon)$. Let $V_j^i = V_j \cap L_{x_i}$. Finally, we lift y_j's to points $\xi_j^i \in V_j^i$.

By the properties of the foliated heat kernel $p_t(x, y)$ on M there exists $N_\varepsilon \in \mathbb{N}$ such that for all $n > N_\varepsilon$

$$\sum_{j=0}^{k} |\mu_{x_n}(V_j^n) - \mu_x(U_j)| \leq \frac{\varepsilon}{2} \text{ and } |\mu_{x_n}(L_{x_n} \setminus (\bigcup_{j=0}^{k} V_j^n)) - \mu_x(L_x \setminus (\bigcup_{j=0}^{k} U_j))| \leq \frac{\varepsilon}{2}.$$

Let $n > N_\varepsilon$. Fix $z \in M$ and define measures

$$\nu = \sum_{j=0}^{k} \mu_x(U_j)\delta_{x_j} + (1 - \sum_{j=0}^{k} \mu_x(U_j))\delta_z$$

and

$$\nu_n = \sum_{j=0}^{k} \mu_{x_n}(V_j^n)\delta_{\xi_j^n} + (1 - \sum_{j=0}^{k} \mu_{x_n}(V_j^n))\delta_z.$$

By Lemma 1.4, measures ν and ν_n satisfy $d_W(\nu, \mu_x) \leq (\mathrm{diam}(M) + 1)\varepsilon$ and $d_W(\nu_n, \mu_{x_n}) \leq (\mathrm{diam}(M) + 1)\varepsilon$. By Lemma 1.3, $d_W(\nu, \nu_n) \leq (3\mathrm{diam}(M) + 2)\varepsilon$. Finally, for every $n > N_\varepsilon$,

$$d_W(\mu_x, \mu_{x_n}) \leq (5\mathrm{diam}(M) + 4)\varepsilon,$$

and $\lim_{i \to \infty} \mathrm{D}_t d(x_i, x) = 0$.

Now, let $\lim_{i \to \infty} \mathrm{D}_t d(x_i, x) = 0$. Suppose that the sequence $(x_i)_{i \in \mathbb{N}}$ does not $\mathrm{D}_0 d$-converge to x. If so, since M is compact, there exists a subsequence $(x_{i_k})_{k \in \mathbb{N}}$ converging to $x' \neq x$ in $\mathrm{D}_0 d$. By the first part of this proof, the subsequence $x_{i_k} \to x'$ in $\mathrm{D}_t d$, as $k \to \infty$. Hence $(x_i)_{i \in \mathbb{N}}$ does not converge to x. Contradiction finishes our proof. $\square$

Corollary 4.1 *M_t is compact and complete.*

Remark 4.1 Due to the properties of the Wasserstein distance of measures, the metric diffusion leaves the set of compact leaves unchanged in the following sense:

Let $(M, \mathscr{F}, g)$ be a compact foliated Riemannian manifold, and let L, L' be two different compact leaves of $\mathscr{F}$. Denote by $\mathrm{D}_t d$ the diffused metric at time $t > 0$. There exists a constant $\eta_{L,L'} > 0$ such that

$$\mathrm{D}_t d(L, L') > \eta_{L,L'} \text{ for all } t \geq 0.$$

This means that any two compact stay in the strictly positive distance greater than $\eta_{L,L'}$. Thus L and L' cannot collapse to the same point, while the collapse is regarded in the set $\mathscr{P}(M)$ of all Borel probability measures equipped with Wasserstein–Hausdorff distance of closed subsets.

One can ask if the metric diffusion be considered for any partition of M by p-dimensional manifolds which does not define a foliation. Observe that we could use an induced measure diffusion along the submanifolds of the partition. In general, the answer is negative, since the topology of M induced by the diffused metric can be different than the original one. The following example illustrates this fact.

Example 4.1 Consider a cylinder $\mathscr{C} = S^1 \times \mathbb{R} \subset \mathbb{R}^3$, where S^1 is placed in XY-plane, and $\mathbb{R}$ is parallel to the Z-axis. For fixed $\tau \in [0, 1)$, we cut the cylinder by a two-dimensional plane π_τ consisting the Y-axis and making with the XY-plane the angle α, with $\sin(\alpha) = \tau$ (Figure 4.1). The set $\pi_\tau \cap \mathscr{C}$ is the ellipse η_τ. Translating η_τ along the Z-axis we get a foliation $\tilde{\mathscr{F}}_\tau$ on $\mathscr{C}$. Dividing $\mathscr{C}$ by $\mathbb{Z}$ in the Z-direction we obtain a compact foliation $\mathscr{F}_\tau$ on $T^2 = S^1 \times S^1$ (Figure 4.2).

Let $M = T^2 \times [0, 1] \subset \mathbb{R}^5$. For each $\tau \in [0, 1)$, we foliate $T^2 \times \{\tau\}$ by $\mathscr{F}_\tau$, while $T^2 \times \{1\}$ is foliated by circles $\{p\} \times S^1 \times \{1\}$. Obviously, there is a singularity along the antipodal circles $U_0 = \{-1\} \times S^1 \times \{1\}$ and $U_1 = \{1\} \times S^1 \times \{1\}$ contained in $T^2 \times \{1\}$, so the partition $\mathscr{P} = \bigcup \mathscr{F}_\tau$ does not define a foliation.

Every set of $\mathscr{P}$ is a circle. Thus, for any $t > 0$ and $x \in M$ one can define a measure $\mathrm{D}_t \delta_x$ as a diffused Dirac mass on a circle of appropriate length. On (M, d),

Fig. 4.1 A foliation $\tilde{\mathscr{F}}_\tau$ of
$S^1 \times \mathbb{R} \times \{\tau\}$

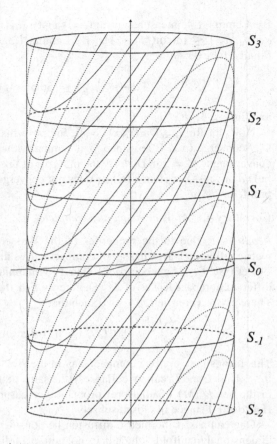

where d is a metric induced from $\mathbb{R}^5$, we define the metric $\mathrm{D}_t d$ diffused along the sets of $\mathscr{P}$ as

$$\mathrm{D}_t d(x, y) = d_W(\mathrm{D}_t \delta_x, \mathrm{D}_t \delta_y).$$

Let us fix $t > 0$ and choose two antipodal points $x_0, x_1 \in S^1$. Let us set $p_0 = (1, x_0, 1)$ and $p_1 = (1, x_1, 1)$. The points p_0 and p_1 divide U_1 into two disjoint arcs $C_0 \ni x_0$ and $C_1 \ni x_1$ of the same length. Let $p_n = (1, x_0, 1 - \frac{1}{n})$, $n > 1$. Obviously $p_n \to p_1$ in d as $n \to \infty$.

Let $A = S^1 \times C_0 \times [0, 1])$ and $B = S^1 \times C_1 \times [0, 1])$. Denote by A_n this connected component of $L_{x_n} \cap A$ which contains p_n. Due to the construction of $\mathscr{P}$,

$$|\mathrm{D}_t \delta_{p_n}(A) - \mathrm{D}_t \delta_{p_n}(B)| = \frac{1}{2} \int_{A_n} p(x_n, y; t)\,dy = \eta_n \xrightarrow[n \to \infty]{} \eta_t > 0.$$

Fig. 4.2 A foliation $\mathscr{F}_\tau$ of $T^2 \times \{\tau\}$ (identify S_0 with S_1)

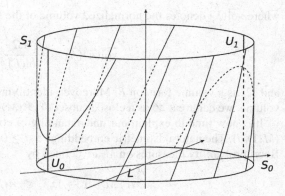

Moreover, $D_t\delta_{p_1}(U_1 \cap A) = D_t\delta_{p_1}(U_1 \cap B) = \frac{1}{2}$, while

$$D_t\delta_{p_n}(A) \xrightarrow[n \to \infty]{} \frac{1}{2} + \eta_t \quad \text{and} \quad D_t\delta_{p_n}(B) \xrightarrow[n \to \infty]{} \frac{1}{2} - \eta_t.$$

Therefore $D_t\delta_{p_n} \xrightarrow[n \to \infty]{} D_t\delta_{p_0}$. Hence, topologies of M induced by d and $D_t d$ are different.

4.2 Metric Diffusion Along Compact Foliations

The classical result says that on a compact manifold M the heat is evenly distributed over M as time tends to infinity. More precisely, let (M, g) be a compact Riemannian manifold. Following [4], we have

Theorem 4.2 *For any $f \in L^2(M)$, the function $D_t f$ converges uniformly, as t goes to infinity, to a harmonic function on M. Since M is compact, the limit function is constant.*

On the other hand, if we focus our attention to the foliated heat diffusion, we can find a number of difficulties with the foliated heat kernel on leaves with different topology and geometry. This even happens if we restrict to a compact foliation. The easier case is a compact foliation with empty bad set. The others can be really confusing. This is not only because of the unbounded volume of leaves but also due to the extrinsic geometry of foliation, that is the geometry of the leaves treated as submanifolds of the foliated manifold.

We begin with the foliation with empty bad set. Due to Theorem 3.1, the leaf space $\mathscr{L}$ of $\mathscr{F}$ is Hausdorff, and hence metrizable. Let $L, L' \in \mathscr{F}$ be two leaves. Define a metric ρ_{vol} in the space of leaves $\mathscr{L}$ by

$$\rho_{\text{vol}}(L, L') = d_W(\overline{\text{vol}}(L), \overline{\text{vol}}(L')),$$

where $\overline{\mathrm{vol}}(F)$ denotes the normalized volume of the leaf F, i.e,

$$\overline{\mathrm{vol}}(F)(A) = \frac{\int_A d\Omega_F}{\mathrm{vol}(F)},$$

and Ω_F is a volume form on F. Moreover, identifying each leaf with its normalized volume, we can treat $\mathscr{L}$ as a closed subset of $\mathscr{P}(M)$.

It is now time to explain our understanding of convergence of the family $M_t = (M, D_t d)$. The natural isometric embeddings $\iota_t, t \geq 0$, of M into the set $\mathscr{P}(M)$ of all Borel probability measures on M are defined by

$$\iota_t : M \ni x \mapsto D_t \delta_x \in \mathscr{P}(M),$$

where δ_x denotes, as before, the Dirac mass concentrated at x. Since M is compact, $(M, D_t d)$ can be treated as a closed subset of $\mathscr{P}(M)$. Thus, for all $t, s \geq 0$ we can study the Wasserstein–Hausdorff distance d_{WH} of M_t and M_s defined by

$$d_{\mathrm{WH}}(M_t, M_s) = \inf\{\varepsilon > 0 : \iota_t(M) \subset N_W(\iota_s(M), \varepsilon) \text{ and } \iota_s(M) \subset N_W(\iota_t(M), \varepsilon)\},$$

where $N_W(A, \eta) = \{\mu \in \mathscr{P}(M) : d_W(\mu, A) < \eta\}$.

Theorem 4.3 *Let $(M, \mathscr{F}, g)$ be a compact foliated manifold carrying a compact foliation with empty bad set. The Wasserstein–Hausdorff limit of $(M)_{t \geq 0}$, $t \to \infty$, is equal to the space $\mathscr{L} = (\mathscr{L}, \rho_{\mathrm{vol}})$ of normalized volumes of leaves.*

Proof Let $\varepsilon > 0$. Since the bad set is empty, the volume of all leaves is commonly bounded by a constant $C > 0$. Hence, there exists $T > 0$ such that for all $x \in M$

$$d_W(\overline{\mathrm{vol}}(L_x), D_t \delta_x) < \varepsilon.$$

Furthermore, $\tilde{\mathscr{L}} \subset N_W(M_t, \varepsilon)$ and $M_t \subset N_W(\tilde{\mathscr{L}}, \varepsilon)$ for all $t > T$. Finally, $d_{\mathrm{WH}}(\tilde{\mathscr{L}}, M_t) \to 0$ as $t \to \infty$. This ends our proof. $\square$

As we have mentioned in the previous chapter, it often happens that the bad set B of a compact foliation is non-empty. In that case, the metric diffusion becomes more complicated, since near the bad set the volume of leaves is unbounded. At fixed time $t_0 > 0$, diffused Dirac measure $D_{t_0} \delta_x$, $x \in M$, can be (in the Wasserstein metric)

1. close to the normalized volume $\overline{\mathrm{vol}}(L_x)$,
2. still not far from a normalized volume of a certain leaf $L \subset B$ (since the leaves of the good set G of $\mathscr{F}$ are more and more tangent, as we approach to the bad set),
3. somewhere in between the normalized volumes of the leaves of G and of B.

Moreover, due to Theorem 4.2, any δ_x, as $t \to \infty$, points toward the point $\overline{\mathrm{vol}}(L_x)$.

Let us assume that M is carrying a compact foliation $\mathscr{F}$ with non-empty bad set. We now ask if $(M_t)_{t \in \mathbb{R}}$, treated as a subset of $\mathscr{P}(M)$, converge in the Wasserstein–Hausdorff distance to a closed subset of $\mathscr{P}(M)$? More general, for compact foliations with non-empty Epstein hierarchy, we seek for necessary condition of Wasserstein–Hausdorff convergence of $(M_t)_{t \geq 0}$.

Before we present the main results of this book, let us study the examples demonstrating that in general there is no convergence at all. Let us begin with an example of a compact foliated space carrying compact foliation with a non-empty bad set.

Following [2], one can extend the definition of foliated heat kernel and foliated heat diffusion to the foliated spaces. This allows us to define the *metric diffusion along foliated space* $(X, d, \mathscr{F})$ by the formula

$$D_t d(x, y) = d_W(D_t \delta_x, D_t \delta_y).$$

The only difference is that the metric d on X has no connection with the transversal structure of $\mathscr{F}$.

Let $X = T^2 \times (\{\frac{1}{i}, i \in \mathbb{N}\} \cup \{0\})$, where $T^2 = S^1 \times S^1 \subset \mathbb{C}^2$ denotes the two-dimensional torus. Let us embed X in $\mathbb{R}^5$ and induce the metric d on X from the Euclidean metric in $\mathbb{R}^5$.

Let $T_n = T^2 \times \{\frac{1}{n}\}$, and $T_0 = T^2 \times \{0\}$. Foliate T_n by a foliation $\mathscr{F}_n$ consisting of closed curves $[0, 2\pi] \ni t \mapsto (ae^{it}, be^{if(n)t}, \frac{1}{n})$, where $a, b \in S^1$, and $f : \mathbb{N} \to \mathbb{N}$ is defined as follows:

Set $f(1) = 1$. Let $n \in \mathbb{N}$, and let $\tau_n > 0$ be the smallest time such that

$$d_W(D_t \delta_x, \overline{\mathrm{vol}}(L_x)) < \frac{1}{n}$$

for all $x \in \bigcup_{i=1}^n T_i$ and all $t \geq \tau_n$. Let $f(n+1) > f(n)$ be a natural number such that for all $x = (p, \frac{1}{n+1}) \in T_{n+1}$

$$d_W(D_{\tau_n} \delta_x, \overline{\mathrm{vol}}(L_{(p,0)})) \leq \frac{1}{n+1}.$$

We complete the foliation of X foliating T_0 by $\mathscr{F}_0$ consisting of circles $\{s\} \times S^1 \times \{0\}$, $s \in S^1$. We denote this foliation by $\mathscr{F}$, while foliations of T_n induced by $\mathscr{F}$ by $\mathscr{F}_n$. The bad set is equal to T_0, while the good set consists of all other $T_n, n \geq 1$.

Let $Y = \mathrm{cl}(\bigcup_{n \in \mathbb{N} \cup \{0\}} \{\overline{\mathrm{vol}}(L), L \in \mathscr{F}_n\}) \subset \mathscr{P}(X)$ with cl denoting the closure in $\mathscr{P}(X)$. One can discover that the set

$$Y \setminus (\bigcup_{n \in \mathbb{N} \cup \{0\}} \{\overline{\mathrm{vol}}(L), L \in \mathscr{F}_n\})$$

consists only of one point—the normalized volume of T_0. This follows from the fact that all leaves of $\mathscr{F}_n$'s are of the form $t \mapsto (ae^{it}, be^{if(n)t}, \frac{1}{n})$ (the speed on the first S^1 component of T^2 is constant), and because of Lemma 1.3.

Let us consider a sequence $(X_i = (X, D_{\tau_i} d))_{i \in \mathbb{N}}$ (Figure 4.3).

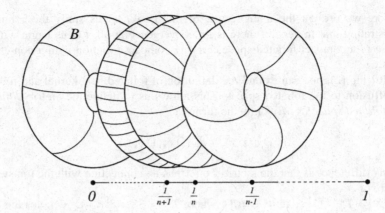

Fig. 4.3 Foliation of X

Lemma 4.1 $(X_i)_{i \in \mathbb{N}}$ *converge in the Wasserstein–Hausdorff distance to Y.*

Proof Let $\varepsilon > 0$ and $n > [\frac{1}{\varepsilon}] + 1$.

1. If $x \in T_m$, $m \le n$, then $d_W(D_{\tau_n}\delta_x, \overline{\text{vol}}(L_x)) < \frac{1}{n} < \varepsilon$.
2. If $x = (p, \frac{1}{m})$, $m > n$, then $d_W(D_{\tau_n}\delta_x, \overline{\text{vol}}(L_{(p,0)})) < \frac{1}{n} < \varepsilon$.

Hence $X_n \subset N(Y, \varepsilon)$ in $\mathscr{P}(X)$ for all $n > [\frac{1}{\varepsilon}] + 1$.

On the other hand, there exists $N > [\frac{1}{\varepsilon}] + 1$ such that for all $n > N$, any leaf L of $\mathscr{F}_n$, and all $x \in T_0$

$$d_W(\overline{\text{vol}}(L), \overline{\text{vol}}(T_0)) < \varepsilon \text{ and } d_W(D_{\tau_n}\delta_x, \overline{\text{vol}}(L_x)) \le \varepsilon.$$

Let $n > N$.

1. If $y = \overline{\text{vol}}(L_0) \in \{\overline{\text{vol}}(L), L \in \mathscr{F}_0\}$, then choosing $x \in L_0$, we have

$$d_W(\overline{\text{vol}}(L_0), D_{\tau_n}\delta_x) \le \frac{1}{N} < \varepsilon.$$

2. For $y = \overline{\text{vol}}(L_0) \in \bigcup_{m>n}\{\overline{\text{vol}}(L), L \in \mathscr{F}_m\}$ there exists $x \in T_N$ such that

$$d_W(y, D_{\tau_n}\delta_x) \le d_W(y, \overline{\text{vol}}(T_0)) + d_W(\overline{\text{vol}}(T_0), D_{\tau_n}\delta_x)$$

$$\le d_W(y, \overline{\text{vol}}(T_0)) + d_W(\overline{\text{vol}}(T_0), \overline{\text{vol}}(L_x))$$

$$+ d_W(\overline{\text{vol}}(L_x), D_{\tau_n}\delta_x) \le 3\varepsilon.$$

3. If $y \in \bigcup_{m=1}^{n}\{\overline{\text{vol}}(L), L \in \mathscr{F}_m\}$, then $y = \overline{\text{vol}}(L)$, and for any $x \in L$

$$d_W(y, D_{\tau_n}\delta_x) < \frac{1}{N} < \varepsilon.$$

4. If $y = \overline{\mathrm{vol}}(T_0)$, then for any point $x \in \mathscr{F}_N$

$$d_W(y, D_{\tau_n}\delta_x) \leq d_W(D_{\tau_n}\delta_x, \overline{\mathrm{vol}}(L_x)) + d_W(\overline{\mathrm{vol}}(L_x), \overline{\mathrm{vol}}(T_0)) \leq 2\varepsilon.$$

Hence, $Y \subset N_W(X_n, 3\varepsilon)$ and $(X_i)_{i \in \mathbb{N}}$ converges to Y.   □

Theorem 4.4 $(X_t = (X, D_t d))_{t \geq 0}$ *does not converge in Wasserstein–Hausdorff topology as* $t \to \infty$.

Proof We will find a sequence $(\theta_i)_{i \in \mathbb{N}}$ for which $(X, D_{\theta_i} d)$ do not converge (while $i \to \infty$) to Y. Let $C = d_W(\{\overline{\mathrm{vol}}(L), L \in \mathscr{F}_0\}, \overline{\mathrm{vol}}(T_0))$. There exists $N \in \mathbb{N}$ such that $d_W(\overline{\mathrm{vol}}(L_x), \overline{\mathrm{vol}}(T_0)) < \frac{C}{16}$ for all $n > N$ and any $x \in T_n$. Let $\varepsilon_0 < \frac{C}{16}$ be a real number for which

$$N(\bigcup_{m=1}^{N} \{\overline{\mathrm{vol}}(L), L \in \mathscr{F}_m\}, \varepsilon_0) \cap N(\{\overline{\mathrm{vol}}(L), L \in \mathscr{F}_0\}, \varepsilon_0) = \emptyset.$$

For every $n > N$, one can choose $\theta_n \in (\tau_n, \tau_{n+1})$ such that for all $x \in T_{n+1}$

$$D_{\theta_n}\delta_x \notin N(\bigcup_{m=1}^{N} \{\overline{\mathrm{vol}}(L), L \in \mathscr{F}_m\}, \varepsilon_0) \cup B(\overline{\mathrm{vol}}(T_0), \frac{C}{16}) \cup N(\{\overline{\mathrm{vol}}(L), L \in (T_0)\}, \varepsilon_0).$$

But on the other hand,

$$Y \subset N(\bigcup_{m=1}^{N} \{\overline{\mathrm{vol}}(L), L \in \mathscr{F}_m\}, \varepsilon_0) \cup N(\overline{\mathrm{vol}}(T_0), \frac{C}{16}) \cup N(\{\overline{\mathrm{vol}}(L), L \in \mathscr{F}_0\}, \varepsilon_0).$$

This implies that the Wasserstein–Hausdorff distance $d_{WH}((X, D_{\theta_n} d), Y) > \varepsilon_0$ for all $n > N$. Hence, the sequence $(X, D_{\theta_n} d)$ does not converge to Y.   □

As a direct conclusion of Theorem 4.4 we obtain that in the case of foliated spaces the result of metric diffusion is unpredictable.

We now restrict to foliated compact Riemannian manifolds. The following example of a compact foliation of dimension one with non-empty bad set and Epstein hierarchy of length one do not converge in Wasserstein–Hausdorff topology.

Following [20], let G be a topological group, while $\gamma : [0, 2\pi] \to G$ a closed curve. Let us define a one-dimensional foliation $\mathscr{F}(\gamma)$ on $S^1 \times G$ filling it with closed curves in such a way that through a point $(t, x) \in S^1 \times G$ passes a curve

$$[0, 2\pi] \ni s \mapsto (s, \gamma(s)\gamma(t)^{-1}x).$$

Leaves of $\mathscr{F}(\gamma)$ are the fibers of a trivial bundle over G with a fiber S^1. Moreover, if G is a Lie group, then $\mathscr{F}(\gamma)$ is a C^r-foliation if only γ is a C^r-curve.

STEP 1 Consider as a Lie group the sphere $S^3 = \{(z, w) \in \mathbb{C}^2 : z\bar{z} + w\bar{w} = 1\}$ with multiplication defined by $(a, b) \cdot (c, d) = (ac - b\bar{d}, ad + b\bar{c})$. For any $\tau \in (0, 1]$, a curve $\gamma_\tau : [0, 2\pi] \to S^3$ as follows:

1. if $\tau = \frac{1}{2n+1} - t, 0 \le t \le \frac{1}{(2n+1)(2n+2)} = a_n, n = 0, 1, 2, \ldots$, then

$$\gamma_\tau(s) = (\sqrt{1 - (\frac{t}{a_n})^2} \, e^{ins}, \frac{t}{a_n} \, e^{ins}), \quad s \in [0, 2\pi];$$

2. if $\tau = \frac{1}{2n} - t, 0 \le t \le \frac{1}{2n(2n+1)} = b_n, n = 1, 2, \ldots$, then

$$\gamma_\tau(s) = (\frac{t}{b_n} \, e^{ins}, \sqrt{1 - (\frac{t}{b_n})^2} \, e^{i(n+1)s}), \quad s \in [0, 2\pi].$$

One can easily check that the family γ_τ is continuous.

We foliate $(0, 1] \times S^1 \times S^3$ setting, for given $\tau \in (0, 1]$, $\mathscr{F}(\gamma_\tau)$ on $\{\tau\} \times S^1 \times S^3$. Directly from the definition of $\mathscr{F}(\gamma_\tau)$, one can see that the length of leaves tends to infinity, and the length of the S^1 component of the vector tangent to a leaf goes to 0 while $\tau \to 0$. Moreover, γ_τ converge tangentially to the left co-sets of closed 1-parameter subgroup $H = \{(e^{is}, 0), s \in [0, 2\pi]\}$. Complementing the foliation of $M = [0, 1] \times S^1 \times S^3$ by adding a foliation of $\{0\} \times S^1 \times S^3$ by leaves of the form

$$\{0\} \times \{t\} \times H \cdot g, \quad g \in S^3, t \in S^1$$

we obtain one-dimensional foliation $\tilde{\mathscr{F}}$ of $[0, 1] \times S^1 \times S^3$ with non-empty bad set $B = \{0\} \times S^1 \times S^3$.

STEP 2 Let $h : [0, 2\pi] \to [0, 2\pi]$ be an increasing C^∞-function with the graph as in Figure 4.4, and equal to the identity in a small neighborhood of 0 and 2π. We will treat h as a function on S^1 with the identity near 1. Next, let $\bar{h} : [0, 1] \times [0, 2\pi] \to$

Fig. 4.4 A modificating function

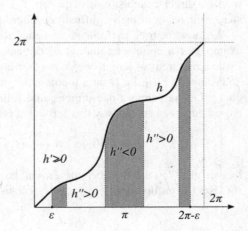

$[0, 2\pi]$ be a smooth homotopy from identity to h, that is $\bar{h}(t, s) = (1 - t)s + th(s)$. Define $\tilde{h} : [0, 1] \times [0, 2\pi] \to [0, 2\pi]$ by the formula

$$\tilde{h}(t, s) = \begin{cases} \bar{h}(2t, s) & \text{for } t \in [0, \frac{1}{2}], \\ \bar{h}(-2t + 2, s) & \text{for } t \in [\frac{1}{2}, 1]. \end{cases}$$

Let $H_n : [0, 1] \times S^1 \times S^3 \to [0, 1] \times S^1 \times S^3$ be given by

$$H_n(\tau, s, x) = \begin{cases} (\tau, \tilde{h}((2n + 1)(2n + 2)\tau - 2n - 1, s), x) \\ \qquad \text{for } (\tau, s, x) \in [\frac{1}{2n+2}, \frac{1}{2n+1}] \times S^1 \times S^3, \\ (\tau, s, x) \qquad \text{otherwise.} \end{cases}$$

Observe that H_n changes $\tilde{\mathscr{F}}$ only on the set $[\frac{1}{2n+2}, \frac{1}{2n+1}] \times S^1 \times S^3$ and leaves it unchanged everywhere else. In other words, it increases the speed in the S^1 direction on two sets, where $h'' > 0$, and decreases it when $h'' < 0$.

STEP 3 Let us modify the foliation $\tilde{\mathscr{F}}$ as follows: For $n_1 = 1$, set $\mathscr{F}_1 = H_1(\tilde{\mathscr{F}})$. Next, choose $\theta_1 > 0$ such that for all $\theta > \theta_1$ and all $p = (\tau, s, x) \in [\frac{1}{2n_1+2}, 1] \times S^1 \times S^3$

$$d_W(D_{\theta_1}\delta_p, \overline{\text{vol}}(L_p)) < \frac{1}{2}.$$

Suppose that we have already chosen $n_k > n_{k-1}$ and $\theta_k > \theta_{k-1}$ such that for foliation $\mathscr{F}_k = H_{n_k} \circ \cdots \circ H_{n_1}(\tilde{\mathscr{F}})$ and all $p = (\tau, s, x) \in [\frac{1}{2(n_k+1)}, 1] \times S^1 \times S^3$

$$d_W(D_{\theta_k}\delta_p, \overline{\text{vol}}(L_p)) < \frac{1}{2^k}.$$

Let us choose $n_{k+1} > n_k$ for which all leaves of $\mathscr{F}_{k+1} = H_{n_{k+1}}(\mathscr{F}_k)$ passing through $p = (\tau, s, x) \in [0, \frac{1}{2n_{k+1}+1}] \times S^1 \times S^3$ satisfies

$$d_W(D_{\theta_k}\delta_p, \overline{\text{vol}}(L_{(0,s,x)})) < \frac{1}{2^{k+1}}.$$

Finally, define the foliation $\mathscr{F}$ as $\lim_{k \to \infty} H_{n_{k+1}}(\mathscr{F}_k)$ and consider M with the Riemannian metric d induced from $\mathbb{R}^7$.

Theorem 4.5 *The family $(M, D_t d)$ does not satisfy the Cauchy condition in Wasserstein–Hausdorff topology. Precisely speaking, there exists $\varepsilon_0 > 0$ such that for any $T > 0$ one can find $\tau, \theta > T$ such that*

$$d_{\text{WH}}(M_\tau, M_\theta) > \varepsilon_0,$$

where $M_a = (M, D_a d)$.

Given $A \subset M$, it is convenient to denote by $\tilde{A}$ the set of the normalized volumes of leaves that passes through A, that is

$$\tilde{A} = \{\overline{\text{vol}}(L_p) : p \in A\}.$$

Proof Denote by B and G the bad set and the good set of $\mathscr{F}$, respectively. Let $C = d_W(\tilde{B}, \text{cl}\tilde{G})$, and $\tau_k = \frac{3+4n_k}{4+12n_k+8n_k^2}$. Observe that

$$\mu = \lim_{k \to \infty} \lim_{t \to \infty} D_t \delta_{(\tau_k, s, x)} \in \text{cl}\tilde{G},$$

with $s \in S^1$ and $x \in T_{\frac{1}{2}} = \{(z, w) \in \mathbb{C}^2 : |z| = |w| = \frac{1}{2}\}$, cannot be obtained as a limit (in Wasserstein metric) of Dirac measure diffused from any point $y \in [0, 1] \setminus \bigcup_{k \in \mathbb{N}} (\frac{1}{2n_k+2}, \frac{1}{2n_k+1}) \times S^1 \times S^3$. This is a direct consequence of the fact that μ has two regions of concentration in direction of S^1 component, while $\lim_{t \to \infty} D_t \delta_y$ is spread evenly. Hence, there exists a constant $\varepsilon_1 \in (0, \frac{C}{8})$ such that

1. $d_W(\mu, \tilde{B}) > 4\varepsilon_1$,
2. for any $k \in \mathbb{N}$ and $p = (\tau, s, x) \in (0, \frac{1}{2n_{k+1}+1}] \times S^1 \times S^3$

$$d_W(D_{\theta_k} \delta_p, \mu) > 2\varepsilon_1.$$

Moreover, for large enough k, we have at $p = (\tau, s, x) \in [\frac{1}{2n_k+2}, \frac{1}{2n_k+1}] \times S^1 \times S^3$

$$d_W(D_{\theta_{k+1}} \delta_p, \text{cl}\tilde{G}) \le 2^{-k} \le \frac{\varepsilon_1}{2} \quad \text{and} \quad d_W(D_{\theta_k} \delta_p, \tilde{B}) < \frac{\varepsilon_1}{2}.$$

Thus, one can find $\theta_k < \lambda_k < \theta_{k+1}$ satisfying for τ_k and any $p \in [0, 1] \times S^1 \times S^3$

$$d_W(D_{\theta_k} \delta_p, D_{\lambda_k} \delta_{(\tau_k, s, x)}) > \frac{\varepsilon_1}{2} = \varepsilon_0.$$

The above inequality implies that $d_{\text{WH}}(M_{\theta_k}, M_{\lambda_k}) > \varepsilon_0$. This ends our proof. □

Theorem 4.5 implies that for a given compact foliation $\mathscr{F}$ on a compact Riemannian manifold (M, g), the family $((M, D_t d))_{t \geq 0}$ does not necessarily converge in Wasserstein–Hausdorff topology as $t \to \infty$. We now formulate the necessary conditions for such convergence.

Let (M, g) be a compact Riemannian manifold carrying a compact foliation $\mathscr{F}$. Let

$$M = B_0 \supset B_1 \supset \cdots \supset B_\beta$$

be the Epstein hierarchy of bad sets with $\text{Eps}(\mathscr{F}) = \beta$. Let $G_i = M \setminus B_i$, $i \le \beta$. For fixed $i \le \beta$, let B_i^j, $j = 1, \ldots, j_i \le \infty$ denote the connected component of B_i. Obviously, $B_i^j \cap B_i^k = \emptyset$ for $j \ne k$. For any saturated set $A \subset M$ we will denote by $N(A, \eta)$ the foliated tubular of A neighborhood of diameter smaller than η.

Theorem 4.6 *If $(M, D_t d)_{t \geq 0}$ converge as $t \to \infty$, then for any B_i^j there exist $\varepsilon_{ij} > 0$ and $C_{ij} > 1$ such that for any $\eta \leq \varepsilon_{ij}$ and leaf $L \subset N(B_i^j, \eta)$ there exists a sequence of leaves $(L_n)_{n \in \mathbb{N}}$ contained in $N(B_i^j, \eta)$ with $L_0 = L$ and $\mathrm{vol}(L_n) \nearrow +\infty$ such that $\mathrm{vol}(L_n) - \mathrm{vol}(L_{n-1}) < C_{ij}$ for all $n \geq 1$.*

Proof Let us suppose conversely that $(M, D_t d)_{t \geq 0}$ converge, and the above condition is not fulfilled. This means that there exists a connected component $B_{i_0}^{j_0}$ of a bad set B_{i_0}, $i_0 \leq \beta$ such that for any $\varepsilon > 0$ and $C > 1$ there exist $\eta_\varepsilon < \varepsilon$ and $L_\varepsilon \subset N(B_{i_0}^{j_0}, \eta_\varepsilon)$ such that for any sequence $(L_\xi) \subset N(B_{i_0}^{j_0}, \eta_\varepsilon)$ with $\mathrm{vol}(L_\xi) \nearrow +\infty$ and $L_0 = L_\varepsilon$ there exists $\xi_0 \in \mathbb{N}$ for which $\mathrm{vol}(L_{\xi_0}) - \mathrm{vol}(L_{\xi_0 - 1}) \geq C$,

STEP 1 We deduce that for any $\varepsilon > 0$ and $C > 1$ one can find $\delta_0 > 0$ and two leaves $F_1, F_2 \subset N(B_{i_0}^{j_0}, \delta_0)$ with $\mathrm{vol}(F_2) - \mathrm{vol}(F_1) \geq C$, such that for any leaf $F \subset N(B_{i_0}^{j_0}, \delta_0)$ either $\mathrm{vol}(F) \leq \mathrm{vol}(F_1)$ or $\mathrm{vol}(F_2) \leq \mathrm{vol}(F)$.

Indeed, let $\varepsilon > 0$, $C > 1$. Let $\delta < \min\{\varepsilon, \inf_k d(N(B_{i_0}^{j_0}, B_{i_0}^k))\}$ and let $L_\delta \subset N(B_{i_0}^{j_0}, \delta)$ be a such leaf that for any sequence of leaves $(L_\xi)_{\xi \in \mathbb{N} \cup \{0\}}$ contained in $N(B_{i_0}^{j_0}, \delta)$ with $\mathrm{vol}(L_\xi) \nearrow +\infty$ there exists $i_\delta \in \mathbb{N}$ for which

$$\mathrm{vol}(L_{i_\delta}) - \mathrm{vol}(L_{i_\delta - 1}) \geq C.$$

Let us fix a sequence $(L_\xi)_{\xi \in \mathbb{N} \cup \{0\}}$, and let $\{i_\alpha\}_{\alpha \in \mathbb{I}} \subset \mathbb{N}$ be the set of all indexes for which $\mathrm{vol}(L_{i_\alpha}) - \mathrm{vol}(L_{i_\alpha - 1}) \geq C$. Let us suppose that for any $\alpha \in \mathbb{I}$ one can find leaves $(F_{i_\alpha}^j)_{j=0}^{k_{i_\alpha}} \subset N(B_{i_0}^{j_0}, \delta)$ with $F_{i_\alpha}^0 = L_{i_\alpha - 1}$, $F_{i_\alpha}^{k_{i_\alpha}} = L_{i_\alpha}$ and satisfying

$$0 < \mathrm{vol}(F_{i_\alpha}^j) - \mathrm{vol}(F_{i_\alpha}^{j-1}) < C$$

for all $j = 1, \ldots, k_{i_\alpha}$. Refilling each gap between leaves $L_{i_\alpha - 1}$ and L_{i_α} by the sequence $(F_{i_\alpha}^j)_{j=0}^{k_{i_\alpha}}$ we obtain, after the re-numeration, a sequence $(L_i)_{i \in \mathbb{N} \cup \{0\}}$ with $L_0 = L_\delta$ for which

$$\mathrm{vol}(L_i) - \mathrm{vol}(L_{i-1}) < C$$

for all $i \in \mathbb{N}$. This gives the contradiction.

STEP 2 For given $C > 1$ and $L \subset N(B_{i_0}^{j_0}, \delta_0)$, let

$$V(L, C) = \{x \in N(B_{i_0}^{j_0}, \delta_0) : \mathrm{vol}(L_x) \geq \mathrm{vol}(L) + C\}$$

$$U(L, C) = N(B_{i_0}^{j_0}, \delta_0) \setminus V(L, C).$$

As before, for given set $A \subset M$, we denote by $\tilde{A}$ the set of all normalized volumes of all leaves passing through A, that is $\tilde{A} = \{\overline{\mathrm{vol}}(L_p), p \in A\}$.

Since $B_{i_0}^{j_0} \cap B_{i_0}^k$ for all $k \neq j_0$, and $B_{i_0}^{j_0} \setminus B_{i_0+1}$ is open and dense in $B_{i_0}^{j_0}$ dense (see Corollary 3.1), there exists a closed connected saturated set $K \subset B_{i_0}^{j_0} \setminus B_{i_0+1}$ satisfying

1. $\text{int}K \neq \emptyset$, where $\int$ is in $B_{i_0}^{j_0}$,
2. $\max_{x \in K} \text{vol}(L_x)$ is bounded,
3. $d_W(\tilde{K}, \text{cl}\tilde{\Sigma}_{\delta_0}) > 0$ where $\Sigma_{\delta_0} = N_W(B_{i_0}^{j_0}, \delta_0)$ and the closure cl is taken in $\mathscr{P}(M)$.

Let $\varepsilon_0 < \frac{1}{16}\min(d_W(\tilde{K}, \text{cl}\tilde{\Sigma}_{\delta_0}), d_W(\tilde{K}, \tilde{G}_{i_0}), \delta_0)$. We can choose $\eta_0 < \varepsilon_0$, $C_0 > 1$, $L_0 \subset G_{i_0} \cap N(K, \eta_0)$, and $\tau_0 > 0$ such that for any $x \in V(L_0, C_0) \cap N(K, \eta_0)$ there exists $y \in K$ such that

$$d_W(D_{\tau_0}\delta_x, \overline{\text{vol}}(L_y)) < \eta_0$$

and for any $x \in U(L_0, C_0)$ the measure $D_{\tau_0}\delta_x \in N_W(\text{cl}\tilde{G}_{i_0}, \eta_0)$.

Indeed, let us suppose that for any $\eta < \varepsilon_0$, $C > 1$, $L \subset G_{i_0} \cap N(K, \eta)$, and $\tau > 0$ we can find a point $x_0 \in V(L, C) \cap N(K, \eta_0)$ such that for any $y \in K$

$$d_W(D_\tau\delta_{x_0}, \overline{\text{vol}}(L_y)) \geq \eta,$$

or there exists $y_0 \in U(L, C)$ for which $D_\tau\delta_{y_0} \notin N_W(\text{cl}\tilde{G}_{i_0}, \eta)$.

The second assertion gives $d_W(D_\tau\delta_x, \tilde{G}_{i_0}) \geq \varepsilon_0$ for all τ, which contradicts with Theorem 4.2.

The first assertion implies that for any fixed $\eta < \varepsilon_0$ and $\tau > 0$ such that for all $x \in K$

$$d_W(D_\tau\delta_x, \overline{\text{vol}}(L_x)) \leq \frac{\eta}{4},$$

we can find a leaf L_x passing through $V(L, C) \cap N(K, \eta)$ with volume as large as we wish. Thus for N's large enough, we can construct a family of leaves $L_{x_N} \subset V(L, N)$, indexed by N's, with $\text{vol}(L_{x_N}) \nearrow +\infty$ as $N \to \infty$, and the sequence $(x_N)_{N \in \mathbb{N}}$ of points of M converging to some point $x_0 \in K$ (because M is compact, and the leaves $L_{x_N}^N$ approach the set $K \subset B_{i_0}^{j_0}$ as $N \to \infty$). On the other hand, for all N

$$d_W(D_\tau\delta_{x_N}, \tilde{K}) > \eta.$$

This contradicts with Theorem 4.1.

STEP 3 Repeating the above arguments, we can construct sequences $(\eta_i)_{i \in \mathbb{N} \cup \{0\}}$, $(C_i)_{i \in \mathbb{N} \cup \{0\}}$ and $(\tau_i)_{i \in \mathbb{N} \cup \{0\}}$ satisfying

$$\eta_i \searrow 0, \quad C_i \nearrow +\infty, \quad \tau_i \nearrow +\infty,$$

and find leaves $L_1, L_2, \ldots$ for which $U(L_i, C_i) \subset U(L_{i+1}, C_{i+1})$, and such that for any $i \in \mathbb{N}$ and $x \in V(L_i, C_i) \cap N(K, \eta_i)$ there exists $y \in K$ such that

$$d_W(D_{\tau_i}\delta_x, \overline{\text{vol}}(L_i)) < \eta_i$$

and for any $x \in U(L_i, C_i)$ the measure $D_{\tau_i}\delta_x \in N_W(\text{cl}\tilde{G}_{i_0}, \eta_i)$.

Let $\theta \in (\tau_i, \tau_{i+1})$, and let

$$A_{i,\theta} = \{\mathrm{D}_{\theta_i}\delta_x : x \in (V(L_i, C_i) \setminus V(L_{i+1}, C_{i+1})) \cap N(K, \eta_i)\}.$$

Since K is a closed subset of $B_{i_0}^{j_0}$ with bounded volume, we can find $\theta_i \in (\tau_i, \tau_{i+1})$ and $p_i \in A_{i,\theta_i}$ such that

$$d_W(\mathrm{D}_{\theta_i}\delta_{p_i}, M_{\tau_i}) > \varepsilon_0.$$

Thus the family $(M, \mathrm{D}_t d)$ does not satisfy the Cauchy condition, and cannot converge in Wasserstein–Hausdorff topology. This contradiction ends our proof. $\qquad \square$

Let us now denote by $N_i(A, \eta)$ the η-neighborhood in B_i of a set $A \subset B_{i+1}$.

Theorem 4.7 *If $(M, \mathrm{D}_t d)_{t \geq 0}$ converge as $t \to \infty$, then for any connected component B_i^j there exist $\varepsilon_{ij} > 0$ such that for any $\eta \leq \varepsilon_{ij}$ and leaf $L \subset N_{i-1}(B_i^j, \eta)$ there exists a sequence of leaves $(L_n)_{n \in \mathbb{N}}$ contained in $N_{i-1}(B_{i-1}, \eta)$ with $L_0 = L$ and $\mathrm{vol}(L_n) \nearrow +\infty$ such that the sequence $(\overline{\mathrm{vol}}(L_n))_{n \in \mathbb{N}}$ converges to some Borel probability measure μ_L with support in B_i^j. Moreover, the support of μ_L is saturated.*

Proof Let us suppose conversely that there exists $B_{i_0}^{j_0}$ that for any $\varepsilon_0 > 0$ there exist $\dot{\eta} < \varepsilon_0$ and $L_\eta \subset N(B_{i_0}^{j_0}, \eta)$ such that for any sequence $(L_i) \subset N_{i-1}(B_{i_0-1}, \eta)$ with $\mathrm{vol}(L_i) \nearrow +\infty$ and $L_0 = L_\eta$, the sequence $(\overline{\mathrm{vol}}(L_i))_{i \in \mathbb{N}}$ does not converge to any measure supported on $B_{i_0}^{j_0}$. Since $\mathscr{P}(M)$ is compact it is enough to suppose that $(\overline{\mathrm{vol}}(L_i))_{i \in \mathbb{N}}$ converges to a measure μ_L which is not supported on $B_{i_0}^{j_0}$. Note that for any $t > 0$ and any $x \in M$ the measure $\mathrm{D}_t\delta_x$ is supported on whole leaf L_x. It follows that there exists $y \in B_{i-1} \cap \mathrm{supp}\mu_L$. Thus, $y \in B_{i_0}^{j_0}$ because $\mathrm{vol}(L_i) \nearrow +\infty$. This gives the contradiction.

To prove that $\mathrm{supp}\mu_L$ is saturated, we suppose conversely, that $\mathrm{supp}\mu_L$ isn't saturated. Thus there exists $x \in \mathrm{supp}\mu_L$ and $y \in L_x$ such that $y \notin \mathrm{supp}\mu_L$. Since $\mathrm{supp}\mu_L$ is closed, there exists $\varepsilon_0 > 0$ such that for all $k \in \mathbb{N}$ one can find $n_k > k$ for which $L_{n_k} \cap B(y, \varepsilon_0) = \emptyset$. Hence, there exists a subsequence $(L_{n_k})_{n_k \in \mathbb{N}}$ such that $\overline{\mathrm{vol}}(L_{n_k}) \not\to \mu_L$ as $k \to \infty$ in Wasserstein distance. Finally, $\overline{\mathrm{vol}}(L_n) \not\to \mu_L$ as $n \to \infty$. This gives the desired contradiction. $\qquad \square$

We can modify the conditions described in Theorems 4.6 and 4.7 to formulate the sufficient condition of convergence for metric diffusion along compact foliations of dimension one with finite Epstein hierarchy.

To begin, let $(M, \mathscr{F}, g)$ be a compact foliated Riemannian manifold carrying a compact foliation of dimension one with Epstein hierarchy

$$M = B_0 \supset B_1 \supset \cdots \supset B_\beta$$

of finite length $\mathrm{Eps}(\mathscr{F}) = \beta$. Let $G_i = M \setminus B_i$, $i \le \beta$. For fixed $i \le \beta$, let B_i^j, $j = 1, \ldots, j_i < \infty$ denote the connected component of B_i. Let K be a connected saturated subset of B_i^j. We define the η-*tiling* based on K in the following way:

First, choose leaves $F_1, \ldots, F_k \in K$ that form 2η net on K. Define a covering $\mathscr{U}$ of the tubular neighborhood $N(K, \eta)$ in M of K by the sets U_i^j with pairwise disjoint interiors and of the form $[a_i^j, b_i^j] \times [-\eta_i^j, \eta_i^j]^q$, $q = \mathrm{codim}\mathscr{F}$, where

1. $i = 1, \ldots, k, j = 1, \ldots, k_i \in \mathbb{N}$,
2. the intervals $[a_i^j, b_i^j] \times \{0\}$, $j = 1, \ldots, k_i$, are contained in F_i, $\bigcup_{i=1}^k [a_i^j, b_i^j] = F_i$, and the open intervals (a_i^j, b_i^j) are pairwise disjoint,
3. the length of $[a_i^j, b_i^j]$'s is smaller than η,
4. the q-dimensional cube $[-\eta_i^j, \eta_i^j]$ is transversal to $\mathscr{F}$ and such that

$$\mathrm{diam}([-\eta_i^j, \eta_i^j]^q) < \eta.$$

The family $\mathscr{U}$ will be called the η-tiling based on K (Figure 4.5).

Theorem 4.8 *If there exist constants $C > 0$ and $\varepsilon_0 > 0$ such that for any B_i^j, any $\varepsilon < \varepsilon_0$, and any $L \subset N(B_i^j, \varepsilon)$ one can find a sequence $(L_i)_{i \in \mathbb{N} \cup \{0\}}$ satisfying*

1. *$L_0 = L$ and for any $i \in \mathbb{N}$ the leaf $L_i \subset N(B_i^j, \varepsilon)$,*
2. *$\mathrm{vol}(L_i) \nearrow \infty$,*
3. *$\mathrm{vol}(L_{i+1}) - \mathrm{vol}(L_i) \le C$,*
4. *the sequence $\overline{\mathrm{vol}}(L_i)$ converges to a measure μ_L and for all $i \ge 0$ we have*

$$d_W(\overline{\mathrm{vol}}(L_i), \mu_L) \le \varepsilon,$$

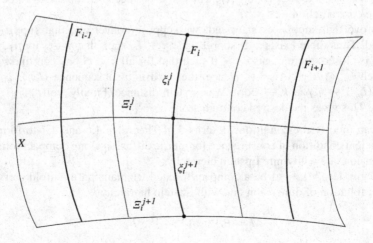

Fig. 4.5 η-Tiling

5. *for any compact saturated set $K \subset B_i^j$ the number of connected components of*
   the set $L_i \cap N(K, \varepsilon)$ is constant for all $i \in \mathbb{N} \cup \{0\}$,

then the family $(M_t)_{t \geq 0}$ satisfies Cauchy condition, i.e., for any $\varepsilon > 0$ there exists
$T > 0$ such that for all $t, s \geq T$

$$d_{\mathrm{WH}}(M_t, M_s) < \varepsilon.$$

Proof Let $0 < \varepsilon < \varepsilon_0$. Denote by Γ_η^τ, $\tau > 0$ the set of all points $x \in M$ for which

$$d_W(\mathrm{D}_\tau \delta_x, \overline{\mathrm{vol}}(L_x)) \leq \frac{\eta}{3 \mathrm{diam}(M) + 2}.$$

Obviously $\Gamma_\varepsilon^t \subset \Gamma_\varepsilon^s$ if only $s > t$. Let $\Delta_\varepsilon = \bigcup_{i=0}^\beta (N_i(\varepsilon) \setminus B_i)$, here $N_i(\varepsilon)$ is a
saturated ε-neighborhood of B_{i+1} in B_i. Let us choose $T > 0$ that

$$M \setminus \Delta_\varepsilon \subset \Gamma_\varepsilon^T. \qquad (4.1)$$

Let $s > t > T$. We now show that $M_s \subset N_W(M_t, \varepsilon)$. Let $x \in M$. Since $\Gamma_\varepsilon^t \subset \Gamma_\varepsilon^s$
then, for any $x \in M \setminus \Delta_\varepsilon$

$$d_W(\mathrm{D}_t \delta_x, \mathrm{D}_s \delta_x) \leq 2\varepsilon.$$

Next, let $x \in \Delta_\varepsilon$. Note that $x \in N(B_i^j, \varepsilon)$ for some i, j. Thus $L_x \subset N(B_i^j, \varepsilon)$, and let
$(L_n)_{n \in \mathbb{N}}$ be a sequence for L_x as assumed. Let $\mathscr{U} = \{U_l^m\}$ be a $\varepsilon/(6\mathrm{diam}(M) + 4)$-
tiling based on $\mathrm{supp}\mu_{L_x}$. Define $U_l = \sum_{j=1}^{k_i} U_l^j$. Suppose that $x \in U_1$. By
Remark 2.2, with $A_i^j = L_x \cap U_i^j$, we can find a leaf L_v and a point $y \in L_v \cap U_1$, such that

$$\sum_{i=1}^k |\mathrm{D}_t \delta_x (L_x \cap U_i) - \mathrm{D}_s \delta_y (L' \cap U_i)| \leq \frac{\varepsilon}{3 \mathrm{diam}(M) + 2}.$$

By Lemma 1.3, $d_W(\mathrm{D}_t \delta_x, \mathrm{D}_s \delta_y) < \varepsilon$, and $d_{\mathrm{WH}}(M_s, M_t) < \varepsilon$. The second inclusion
we prove in the same way. This ends our proof. $\square$

Remark 4.2 The assumption on finiteness of the length of Epstein hierarchy and
finite number of connected components of each B_i is essential. If the length of
the hierarchy is infinite, then no necessarily exists T in the formula (4.1). The
counterexample with the function $\mathbb{N} \ni i \mapsto \min\{\mathrm{vol}(L); L \subset B_i\}$ unbounded can
be easily constructed with the methods described in [20].

Remark 4.3 The set of foliations satisfying conditions 1–5 of Theorem 4.8 is non-
empty, since the foliation described in Example 3.1 satisfies these conditions.

The results of Theorem 4.8 allow us to determine the Gromov–Hausdorff
convergence of the family $(M_t = (M, \mathrm{D}_t d))_{t \geq 0}$. We only recall here the main facts:

Let (X, d_X) and (Y, d_Y) arbitrary compact metric spaces. Set X and Y as

$$d_{\mathrm{GH}}(X, Y) = \inf\{d_H(X, Y)\},$$

where d ranges over all admissible metric on disjoint sum $X \bigsqcup Y$, i.e. d is an extension of d_X and d_Y, while d_H denotes the Hausdorff distance. The number $d_{GH}(X, Y)$ is called the *Gromov–Hausdorff distance* of metric spaces X and Y. It occurs that d_{GH} defines a metric on the class $\mathscr{M}$ of all isometry classes of compact metric spaces.

Remark 4.4 The Gromov–Hausdorff distance can be equivalently defined as follows. For two compact metric space (X, d_X), (Y, d_Y) define

$$\tilde{d}_{GH}(X, Y) = \inf_{Z}\{d_H^Z(f(X), g(Y))\}$$

where (Z, d^Z) ia a metric space, d_H^Z denotes the Hausdorff distance in Z, and $f : X \to Z$, $g : Y \to Z$ are isometric embeddings of X and Y into Z, respectively. Then d_{GH} and $\tilde{d}_{GH}$ are equivalent.

Theorem 4.9 *$\mathscr{M}$ equipped with d_{GH} is complete.*
A proof can be found in [5].

The immediate consequence of Remark 4.4, Theorem 4.8, and Theorem 4.9 is the following.

Corollary 4.2 *Let $(M, \mathscr{F}, g)$ be a compact Riemannian foliated manifold carrying a one-dimensional compact foliation with finite Epstein hierarchy and satisfying the conditions of Theorem 4.8. Then the Gromov–Hausdorff limit $\lim_{t\to\infty}^{\mathrm{GH}} M_t$ exists.*

Proof Due to Remark 4.4, $d_{\mathrm{GH}}(M_t, M_s) \leq d_{\mathrm{WH}}(M_t, M_s)$. Hence $(M_t)_{t\geq 0}$ satisfies the Cauchy condition. By Theorem 4.9, $(\mathscr{M}, d_{\mathrm{GH}})$ is complete. Thus the limit $\lim_{t\to\infty}^{\mathrm{GH}}(M, \mathrm{D}_t d)$ exists.   □

Chapter 5
Metric Diffusion for Non-compact Foliations: Remarks

We begin with the example of a two-dimensional foliation of codimension one having dense leaves with very complicated geometry and dynamics.

Example 5.1 ([21]) Let $\tilde{M} = D^2 \times S^1$ be the solid torus. Let $\phi : \tilde{M} \to \tilde{M}$ be defined by

$$\phi(z, w) = (\frac{1}{2}w + \frac{1}{4}z, w^2)$$

where $|z| \leq 1$ and $|w| = 1$. $M_0 = \tilde{M} \setminus \text{Int}\phi(\tilde{M})$ is a compact 3-manifold with boundary $\partial M_0 = N_0 \cup N_1$, with each N_i diffeomorphic to the two-dimensional torus T^2. M_0 can be foliated in a very natural manner by the foliation $\mathscr{F}_0$ consisting of surfaces $w = \text{const}$.

Identifying in M_0 the components N_0 and N_1 via ϕ we obtain the compact manifold without boundary M carrying the foliation $\mathscr{F}$ induced on M from $\mathscr{F}_0$. $\mathscr{F}$ is called the *Hirsch foliation*. Each leaf of $\mathscr{F}$ is built of the "panties" (see Figure 5.1) which are topologically equivalent to the disk with two holes.

$\mathscr{F}$ has exponential growth, each leaf is everywhere dense in M and of the form of Cantor tree or blossoming Cantor tree (see Figure 5.2). Moreover, $\mathscr{F}$ contains a resilient leaf, and hence an exceptional minimal set. It also has positive entropy (see [21] or [3] for details).

On the other hand, each leaf of the Hirsch foliation has the Cantor set of ends. Thus, the form of the heat kernel for a leaf of $\mathscr{F}$ is unknown. The geometry of a leaf can have a great influence on metric diffusion, since some parts of a leaf can transmit heat better than others. Of course, on each leaf the heat will drift to its ends.

We suspect that even for Hirsch foliation, which is of codimension one, there is no chance of collapsing a single leaf to a point while metric diffuses in time. Thus we formulate an open question:

© The Author(s) 2017
S.M. Walczak, *Metric Diffusion Along Foliations*, SpringerBriefs in Mathematics,
DOI 10.1007/978-3-319-57517-9_5

Fig. 5.1 The "panties"

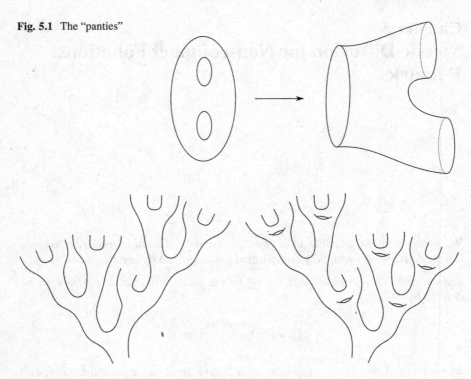

Fig. 5.2 Cantor tree (*left*) and blossoming Cantor tree (*right*)

Let $\mathscr{F}$ be a codimension one foliation on a compact Riemannian manifold (M, g). Is every non-compact leaf of $\mathscr{F}$ collapsing, in Wasserstein–Hausdorff topology, to the point while metric diffuses in time?

The above example demonstrates a number of serious difficulties in the case of non-compact foliations. In next considerations we present some partial results on metric diffusion along one-dimensional foliations having non-compact leaf.

Let us consider a cylinder $\mathscr{C} = S^1 \times [0, 1]$ foliated by circles $L_0 = S^1 \times 0$, $L_1 = S^1 \times 1$ and real lines unwinding from $L - 0$ and accumulating on L_1 (see Figure 5.3) and equipped with the product metric d. There are two possibilities, that the leaves rewind with no change (Kronecker foliation) or with the change (Reeb foliation) of the direction. Let us denote by $\mathscr{F}_K$ and $\mathscr{F}_R$ the Kronecker and the Reeb foliations, respectively.

Let $n \in \mathbb{N}$ be a fixed number and let $A_i = \bigcup_{k \in \mathbb{Z}} [k + \frac{i}{n}, k + \frac{i+1}{n}]$. Let us denote by $p_t(x, y)$ the heat kernel on real line.

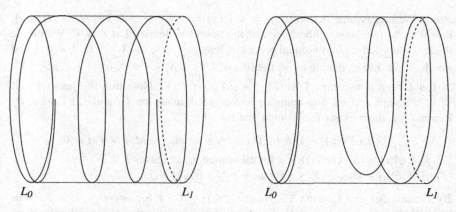

Fig. 5.3 A linear foliation of the cylinder with two compact leaves (*left*: Kronecker foliation; *right*: Reeb foliation)

Lemma 5.1 *For any $x \in \mathbb{R}$ and $0 \le i \le n-1$, we have $\int_{A_i} p_t(x,y)dx \to \frac{1}{n}$ as $t \to \infty$.*

Proof Let $x \in \mathbb{R}$. First, observe that there exists a bounded interval D such that

$$\int_{A_i \setminus D} p_t(x,y)dx \le \int_{A_j} p_t(x,y)dx$$

for all $i,j \in \{0,\ldots,n-1\}$ and $t > 0$. Set $i \in \{0,\ldots,n-1\}$. We have

$$n \cdot \int_{A_i \setminus D} p_t(x,y)dx \le \sum_{j=0}^{n-1} \int_{A_j} p_t(x,y)dx.$$

Thus $\int_{A_i \setminus D} p_t(x,y) \le \frac{1}{n}$, and $\int_{A_i} p_t(x,y) \le \frac{1}{n} - \int_D p_t(x,y)dx$. Since $\int_D p_t(x,y)dx \to 0$ as $t \to \infty$, then $\int_{A_i} p_t(x,y) \le \frac{1}{n}$. Conversely, we have

$$n \cdot \int_{A_i} p_t(x,y)dx \ge \sum_{j=0}^{n-1} \int_{A_j \setminus D} p_t(x,y)dx = 1 - \int_D p_t(x,y)dx.$$

This gives us the proof.   □

Theorem 5.1 *The family $(\mathscr{C}_t = (\mathscr{C}, \mathrm{D}_t d))_{t \ge 0}$ with metric d diffused along $\mathscr{F} = \mathscr{F}_K$ or $\mathscr{F} = \mathscr{F}_R$, satisfies in the Wasserstein–Hausdorff topology, the Cauchy condition, i.e.,*

$$\forall_{\varepsilon > 0} \quad \exists_{T > 0} \quad \forall_{s > t > T} \quad d_{\mathrm{WH}}(\mathscr{C}_s, \mathscr{C}_t) < \varepsilon.$$

Proof We can assume that $\text{vol}(L_0) = \text{vol}(L_1) = 1$. Let $\varepsilon > 0$. Let $N_i = N(L_i, \varepsilon)$, $i = 0, 1$, be tubular neighborhoods of the boundary leaves. Let $n > 0$. We divide the boundary leaves into n equal pieces of length $\frac{1}{n} < \varepsilon$, say $U_j^i \subset L_i$, $i = 0, 1$ and $j = 1, \ldots, n$. Let us denote by V_j^i the lifts of U_j^i's to N_i's, $i = 0, 1$.

Let L be a non-compact leaf. Set $A_i^j(L) = L \cap V_i$. Note that the geometry of $\mathscr{C}$ is bounded and all non-compact leaves accumulate on L_0 and L_1. Hence, by Lemma 5.1, there exists $T > 0$ such that for all $t > T$

(1) $\sum_{j,k=1}^{n} |D_t \delta_x(A_i^j(L_x)) - D_t \delta_x(A_i^k(L_x))| \leq \frac{\varepsilon}{2}$ for all $x \in M, t > T, i = 0, 1$,
(2) $D_t \delta_x(A_i^j(L \setminus (N_0 \cup N_1))) \leq \varepsilon$ for all non-compact leaves $L \in \mathscr{F}$,
(3) $|D_t \delta_x(N_0) - D_t \delta_x(N_1)| \leq \varepsilon$ for all $x \in M \setminus (N_0 \cup N_1)$.

By Lemma 5.1 and Lemma 1.3, $d_{\text{WH}}(\mathscr{C}_s, \mathscr{C}_t) \leq C \cdot \varepsilon$ for any $s > t > T$. Thus $\mathscr{C}_s \subset N_{\text{W}}(\mathscr{C}_t, C \cdot \varepsilon)$ and $\mathscr{C}_t \subset N_{\text{W}}(\mathscr{C}_s, C \cdot \varepsilon)$, for some constant C depending only on M and d.   □

Denote by I the interval of length $l = d_{\text{W}}(\overline{\text{vol}}(L_0), \overline{\text{vol}}(L_1))$, where $\overline{\text{vol}}(L)$ is the normalized volume of a leaf L.

Corollary 5.1 $d_{\text{GH}}(\mathscr{C}_t, I) \to 0$ as $t \to \infty$.

Proof This is the direct consequence of Theorem 5.1 and Theorem 4.9.

Let $T^2 = S^1 \times S^1$ be the two-dimensional torus carrying a one-dimensional foliation $\mathscr{F}$ with at least one compact leaf.

Corollary 5.2 *The Wasserstein–Hausdorff limit of the diffused metric along $\mathscr{F}$ is isometric to a circle.*

Proof If all leaves are compact, then the result is the direct consequence of Theorem 4.3. If at least one leaf is non-compact, then $\mathscr{F}$ consists of a finite number of Reeb components, a countable number of Kronecker components and circles. The result follows directly form Theorem 4.3 and Theorem 5.1.

References

1. Bogachev, V.I.: Measure Theory. Springer, New York (2007)
2. Candel, A.: The harmonic measures of Lucy Garnett. Adv. Math. **176**(2), 187–247 (2003)
3. Candel, A., Conlon, L.: Foliations I & II. American Mathematical Society, Providence (2001 & 2003)
4. Chavel, I.: Eigenvalues in Riemannian Geometry. Academic Press, New York (1984)
5. Chi, D.-P., Yun, G.: Gromov–Hausdorff Topology and Its Applications to Riemannian Manifolds. Seoul Nat. Univ., Seoul (1998)
6. Engelking, R.: General Topology. Heldermann, Berlin (1989)
7. Edwards, R., Millett, K., Sullivan, D.: Foliations with all leaves compact. Topology **16**, 13–32 (1977)
8. Epstein, D.B.A.: Periodic flows on 3-manifolds. Ann. Math. **95**, 66–82 (1972)
9. Epstein, D.B.A.: Foliations with all leaves compact. Ann. Inst. Fourier Grenoble **26**, 265–282 (1976)
10. Epstein, D.B.A., Vogt, E.: A counterexample to the periodic orbit conjecture in codimension 3. Ann. Math. **108**(3), 539–552 (1978)
11. Garnett, L.: Foliations, the Ergodic Theorem and Brownian Motions. J. Funct. Anal. **51**(3), 285–311 (1983)
12. Graf, S., Mauldin, R.D.: The classification of disintegrations of measures. Contemp. Math. **94**, 147–158 (1989)
13. Moerdijk, I., Mrcun, J.: Introduction to Foliations and Lie Groupoids. Cambridge University Press, Cambridge (2003)
14. Reeb, G.: Sur certaines properiétés topologiques des variétés feuilletées. Actual scient. ind. **1183**, 93–154 (1952)
15. Rudin, W.: Real and Complex Analysis. Tata McGraw-Hill, New Delhi (1987)
16. Sullivan, D.: A counterexample to the periodic orbit conjecture. Publ. Math. de l'IHES **46**(1), 5–14 (1976)
17. Tamura, I.: Topology of Foliations: An Introduction. American Mathematical Society, Providence (1992)
18. Villani, C.: Optimal Transport, Old and New. Grundlehren der mathematischen Wissenschaften, vol. 338. Springer, New York (2009)
19. Villani, C.: Topics in Optimal Transportation. American Mathematical Society, Providence (2003)
20. Vogt, E.: A periodic flow with infinite Epstein hierarchy. Manuscripta Math. **22**, 403–412 (1977)

© The Author(s) 2017
S.M. Walczak, *Metric Diffusion Along Foliations*, SpringerBriefs in Mathematics,
DOI 10.1007/978-3-319-57517-9

21. Walczak, P.: Dynamics of Foliations, Groups and Pseudogroups. Birkhäuser, Boston (2004)
22. Walczak, Sz.: Warped compact foliations. Ann. Pol. Math. **94**, 231–243 (2008)
23. Walczak, Sz.: Hausdorff leaf spaces for foliations of codimension one. J. Math. Soc. Jpn. **63**(2), 473–502 (2011)

Index

© The Author(s) 2017
S.M. Walczak, *Metric Diffusion Along Foliations*, SpringerBriefs in Mathematics,
DOI 10.1007/978-3-319-57517-9

Printed in the United States
By Bookmasters